Fieldwork Geography

JANE FERRETTI *and* JANET BOYCE

CAMBRIDGE UNIVERSITY PRESS

Cambridge
London New York New Rochelle
Melbourne Sydney

Acknowledgements
We would like to thank Mrs Christine Naylor for typing the manuscripts and Melvyn Jones for permission to use fig. 13.1 (model of a suburbanised village devised by Vivian Thomas).

Published by the Press Syndicate of the University of Cambridge
The Pitt Building, Trumpington Street, Cambridge CB2 1RP
32 East 57th Street, New York, NY 10022, USA
296 Beaconsfield Parade, Middle Park, Melbourne 3206, Australia

First published 1984

Printed in Great Britain at the University Printing House, Cambridge

ISBN 0 521 27300 5

SE

Contents

Introduction

This book is aimed largely at VIth form students who are required to submit a fieldwork project as part of their examination and assessment. It may also be useful for geography teachers and students in colleges of education who are looking for ideas for fieldwork. Most of the chapters are designed for independent use by students but the final three chapters may need some additional assistance at least in the planning stages.

Fourteen topics have been selected and these include aspects of both physical and human geography. All the examples in the book have been carried out by the authors with groups of students. The topics are arranged to show a progression from easier to more difficult methods of data collection and analysis, and in some of the later chapters it is necessary to refer to methods described in earlier chapters.

All the studies are based on the method of geographical enquiry outlined below which adopts a hypothesis testing approach.

1 Select a topic for study and identify a problem or idea which you wish to study.
2 Set a hypothesis.
3 Decide what data is required and how to collect it.
4 Collect and record data.
5 Process and present your data.
6 Analyse your results
7 Accept or reject your hypothesis
8 If you wish, make further investigations as a result of your findings

Wherever possible metric units have been used but in some cases the use of imperial units is more suitable, for example when discussing distances between towns which are still measured in miles.

Two statistical tests have been included; these are Spearman's rank correlation (r_s) and the chi-square test (χ^2). Although worked examples have been provided in the text it is assumed that students will have a working knowledge of these statistical techniques before attempting to use them in fieldwork studies.

In each chapter suggestions are given for analysing and drawing conclusions, but clearly this must be related to the particular data collected. Nevertheless it must be emphasised that the discussion of the results is a very important part of any fieldwork study. Ideas for further studies are also given at the end of each chapter. It is hoped that the suggestions for fieldwork and the further studies will be adapted and extended to fit particular interests and locations.

Where suitable locations for study are included it must be emphasised that these are areas chosen simply because they fit the requirements well. By following the authors' guidelines it will be possible to choose numerous locations for each study.

Fieldwork is normally done 'in the field'. However, certain investigations, which can still be classified as fieldwork, can be carried out without the need to travel. The topic in chapter 12 is an example and may be suitable where financial considerations or physical handicap prevent outside fieldwork.

1 A study of agricultural land use

Suitable for group or individual study

Topic for study

Land use is influenced by physical, economic and social factors, as illustrated in fig. 1.1. Studies of small areas can increase the understanding of land use patterns, although it will rarely, if ever, be possible to find a complete explanation of farmers' decisions.

The exercise described here concentrates on the influence of physical factors on land use patterns; however, some ideas for looking at the influence of economic factors are given in the final section of the chapter.

Possible hypothesis

Agricultural land use in (the chosen area) is related to altitude, aspect and surface geology.

Location

It is important to choose the area of study very carefully. The area should be between 6 and 8 square kilometres in size and must contain a variety of land use, such as arable land, grassland, rough pasture and woodland. The area should also show variations in altitude, aspect and geology. You can find out whether there are variations in altitude and aspect by studying the 1 : 25,000 OS map for the area. Similarly variations in geology can be seen on the 1 : 25,000 geological sheet (solid and drift). Very complex geological patterns should be avoided because this makes sampling and analysis difficult. The only way to be certain that the land use varies is to visit the area before beginning the study, and it is essential to do this.

Suitable locations include:

1 An escarpment with a north or south facing scarp slope, e.g. Wenlock Edge, Shropshire.
2 A west–east valley in an upland area, e.g. Langdale Valley, Lake District; Edale Valley, Peak District.

You must have access to the whole area you wish to study, so make sure there are public footpaths which will allow you to walk round and through the area you have chosen.

Data collection

First prepare a simple but accurate base map of the area using a 1 : 10,000 OS map. You may be able to use a copy of the map itself or you may have to produce your own map showing field boundaries, contour lines, and other features such as rivers and settlements. Make two copies of this base map.

Then visit the study area, walk all around it and plot the land use of the whole area on one copy of the base map. Use a simple classification system such as

arable land	rough pasture
permanent pasture	marsh
ley grass	horticulture or orchard
woodland	other (e.g. private gardens)

It is essential to match the fields on the ground with those on

1.1 Factors influencing land use

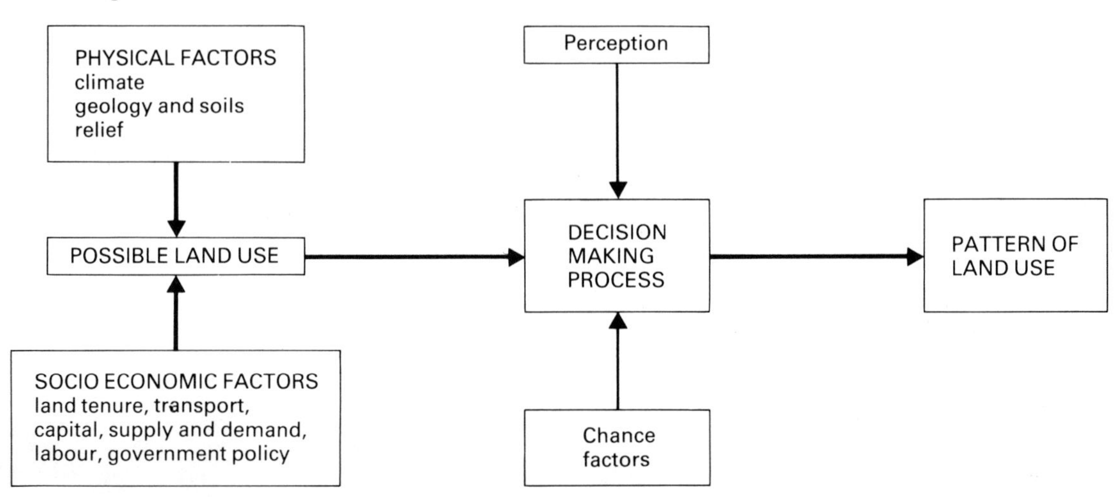

the map. Mistakes can be avoided by finding an obvious landmark from which to start plotting the land use, such as a bridge, a road junction, or a farmhouse with a name. Then try to find a high point from which you have a view over all or part of the area from where you can plot the land use.

Various problems may arise. First it may be difficult to classify land use, for example there may be an area of marshland within a field of permanent pasture. In some cases you may be able to plot the marsh onto the map, alternatively it may be better to classify the field under the major land use. A pair of binoculars is useful to ascertain land use of more distant fields, but in many cases you will have to walk up to the field in order to identify the crop. If you are unsure about how to recognise different land use types, then ask for advice before you start your survey.

A second problem you may come across is that some field boundaries may have been altered. Farmers sometimes construct new fences or more frequently, remove boundaries and this will change the size and shape of the field and means the land will not 'match' the map. However, it is usually quite easy to identify new boundaries and also to see the line of previous ones. The base map should of course be changed to correspond with the existing field pattern.

A good time of year to carry out this study is during the summer when the crops can be recognised more easily.

Processing data

First make a neat and accurate land use map of the area studied using colours and the second copy of the base map. Having mapped the land use pattern it will probably be possible to make some generalisations about the occurrence of different land uses in relation to altitude, aspect and geology. However, in order to obtain *evidence* to test your hypothesis a sample must be taken.

Take a sample of at least 100 points. The larger your sample, the stronger your evidence will be. Use one of the two methods shown below.

Random sampling

Draw a 10 × 10 grid to cover your base map as shown in fig. 1.2a. This may be drawn onto the base map itself or onto tracing paper which should then be attached to the base map as an overlay. Then choose 100 four-figure random numbers, from random number tables (Appendix B) and read these points off from the grid.

Systematic sampling

Draw a 9 × 9 grid to cover your base map as shown in fig. 1.2b. Again this may be drawn onto the base map itself or onto a tracing overlay. Take each intersection as a sample point; this provides 100 points. This method is quicker and easier but could over emphasise a linear feature.

At each sample point note the land use, altitude, aspect and geology.

The land use must be recorded using the categories selected for your fieldwork. Remember to include a column for 'other' on your recording sheet. Your sample may land in a river!

The altitude should be recorded in classes (e.g. 0–30 m, 31–61 m, 62–91 m, using OS metricated contour intervals). You should have a minimum of three and a maximum of six altitude classes.

The aspect should also be recorded in classes. Find the

1.2a Random sampling

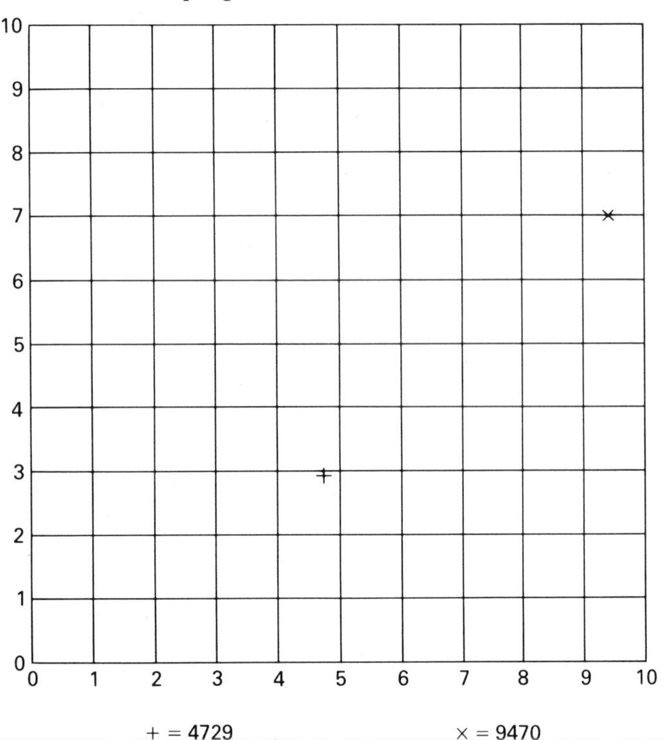

1.2b Systematic sampling (this grid provides 100 sample points)

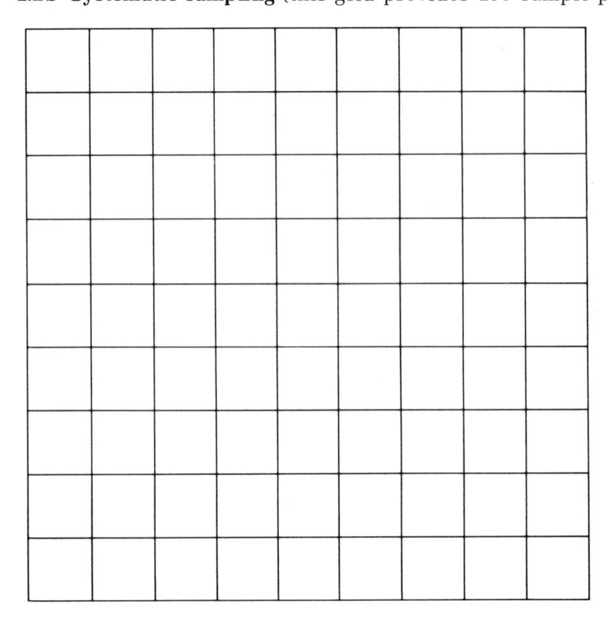

1.3a Reading aspect from a map

N

sample point

107

sample point

76

line drawn at right angles to contours shows aspect SW

line drawn at right angles to contours shows aspect SSE

1.3b Recording aspect in classes

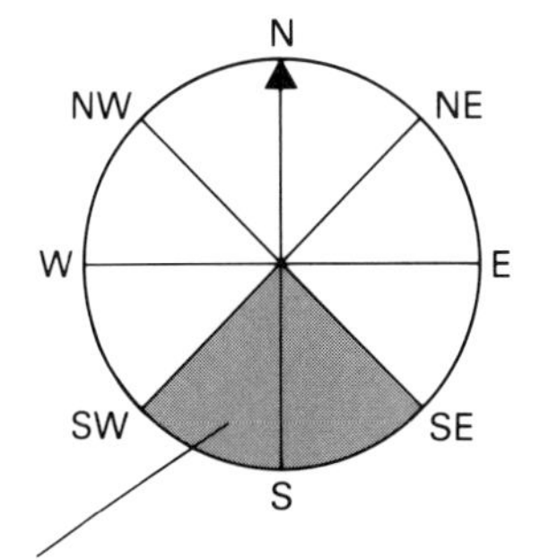

readings falling within this area should be recorded in SE–SW class

direction the slope faces by looking at the contour lines (see fig. 1.3a) then record in classes NW–NE; NE–SE; SE–SW; SW–NW and nil (see fig. 1.3b). Flat or very gently sloping land has no aspect and therefore must be recorded as nil.

The geology for each sample point must be recorded by referring to the Institute of Geological Sciences geological sheet (solid and drift), scale 1 : 25,000.

Record all the information for each sample on a recording sheet (fig. 1.4) then total your samples and complete a summary sheet (fig. 1.5). Check your summary sheet carefully to make sure that you have all sample readings for each variable, altitude, aspect and geology. There should be the same number of readings for each variable.

1.4 Recording sheet

	Land use					Altitude				Aspect					Geology		
Sample point	A	PP	W	RP	O	under 152 m	152–182 m	183–212 m	over 212 m	NW–NE	NE–SE	SE–SW	SW–NW	NIL	BC	S	L
1	✓						✓					✓			✓		
2		✓				✓								✓		✓	
3		✓					✓							✓		✓	
4	✓					✓						✓			✓		
5			✓					✓			✓					✓	
6 etc.					✓	✓								✓		✓	

A = Arable
PP = Permanent pasture
W = Woodland
RP = Rough pasture
O = Other
BC = Boulder clay
S = Shale
L = Limestone

1.5 Summary sheet

These results were collected from an area of Wenlock Edge, Shropshire

	Altitude			Aspect			Geology		
Land use	under 152 m	152–182 m	over 182 m	NW–NE	SE–SW	NIL	BC	S	L
Arable	24	26	6	22	6	28	46	4	6
Permanent pasture	24	12	2	12	2	24	28	10	
Woodland			4	4				4	
Other	2					2		2	
Total	50	38	12	38	8	54	74	20	6
			100			100			100

1.6 Percentage of land use in different altitude classes

	Total	Under 152 m		152–182 m		Over 182 m	
		Number	%	Number	%	Number	%
Arable	56	24	42.9	26	46.4	6	10.7
Permanent pasture	38	24	63.2	12	31.6	2	5.2
Woodland	4	0	0	0	0	4	100

1.7 Cumulative frequency graph showing relationship between land use and altitude

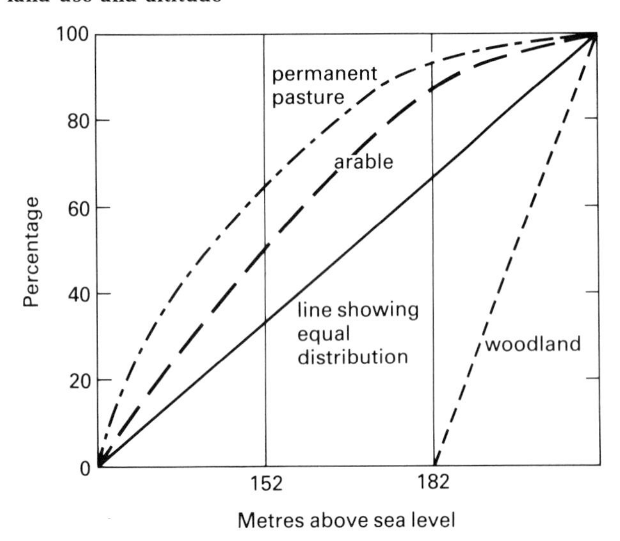

Presentation of results

The data in the summary sheet provide the basis for your analysis. You should present information about land use in relation to each of the variables mentioned in your hypothesis (altitude, aspect and geology). There are several different methods you could use and three possibilities are shown below. These illustrate work in Wenlock Edge, Shropshire, summarised in fig. 1.5.

Construct a cumulative frequency graph to show the relationship between land use and altitude.

Calculate the percentage of each land use which falls into each altitude class and show this in a table (fig. 1.6). Then use this information to construct a cumulative frequency graph (fig. 1.7). For example, 42.9% of the arable land is below 152 metres, an additional 46.4% lies between 152 and 182 metres (total 89.3%). The final 10.7% is over 182 metres (total 100%). All curves, except the line of equal distribution, should be drawn by eye, not with a ruler.

Construct divided proportional circles to show the relationship between land use and aspect (fig. 1.8).

The different sizes of circles represent the amount of land given over to a particular land use, and the divisions show the percentage in each aspect class. To calculate the radius of each circle you must first find the square root of the number of observations for each type of land use, for example, 56 observations for arable land: $\sqrt{56} = 7.48$. Use the square roots to work out a suitable scale for the circles. It is essential to use the square root in order that the areas of the circles are proportional.

When you have drawn the circles subdivide them to show the percentage of observations in each aspect class, using data from your summary table (fig. 1.6). For example, for a total of 56 observations for arable land then each observation will be represented by $360° \div 56 = 6.4°$.

Finally draw a proportional circle that shows the proportions of the whole sample within each aspect class. This should be used for comparison.

Construct a divided column graph to show how much of the area consists of the different rock types, and what the land use is in each of these areas (fig. 1.9).

1.8 Divided proportional circles showing relationship between land use and aspect

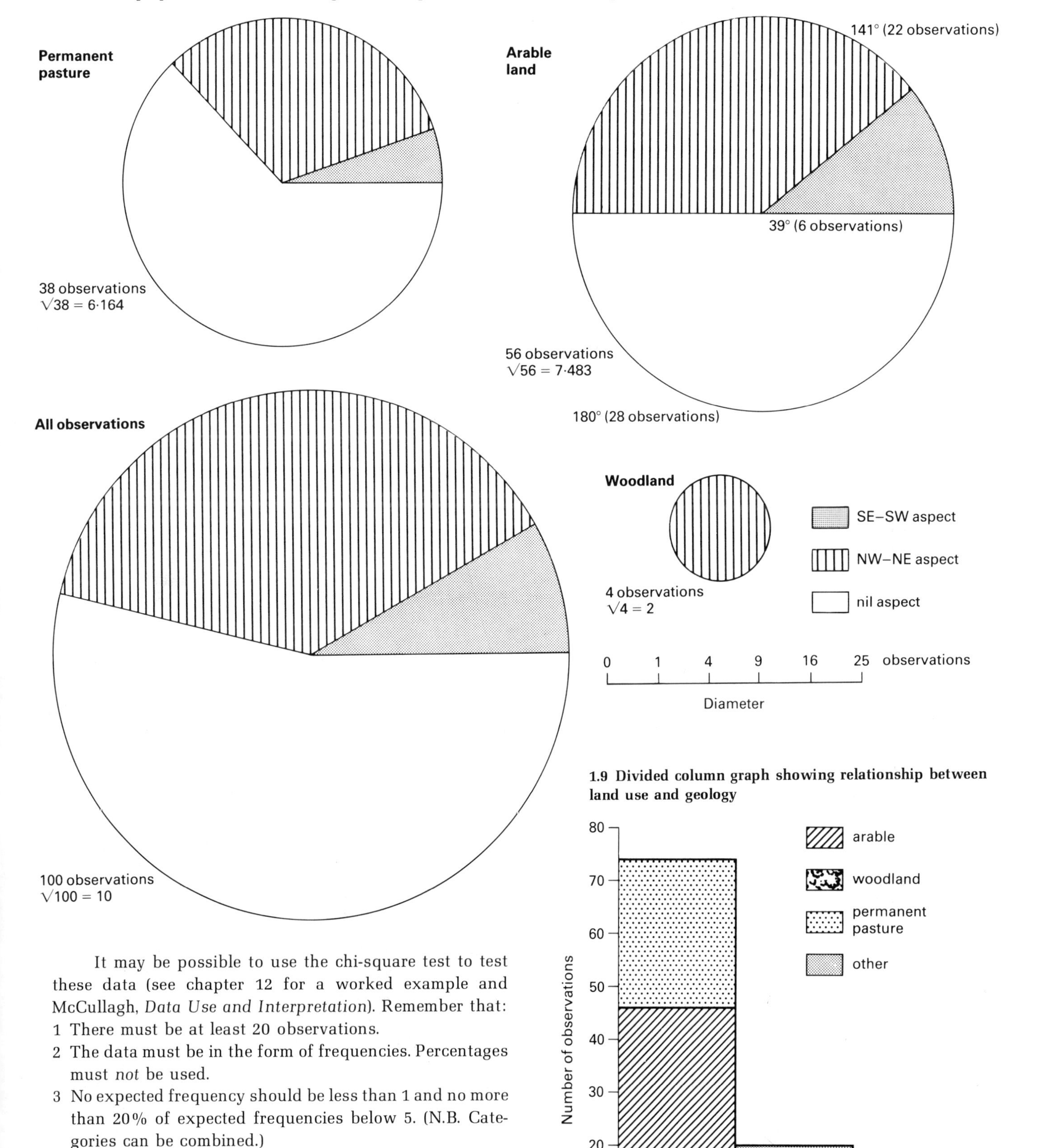

1.9 Divided column graph showing relationship between land use and geology

It may be possible to use the chi-square test to test these data (see chapter 12 for a worked example and McCullagh, *Data Use and Interpretation*). Remember that:

1 There must be at least 20 observations.

2 The data must be in the form of frequencies. Percentages must *not* be used.

3 No expected frequency should be less than 1 and no more than 20% of expected frequencies below 5. (N.B. Categories can be combined.)

4 The observations must be independent.

$$\chi^2 = \sum \frac{(O-E)^2}{E}$$

where O = observed frequency
E = expected frequency

1.10 Cross-section through Wenlock Edge, drawn to half scale from 1:10000 base map

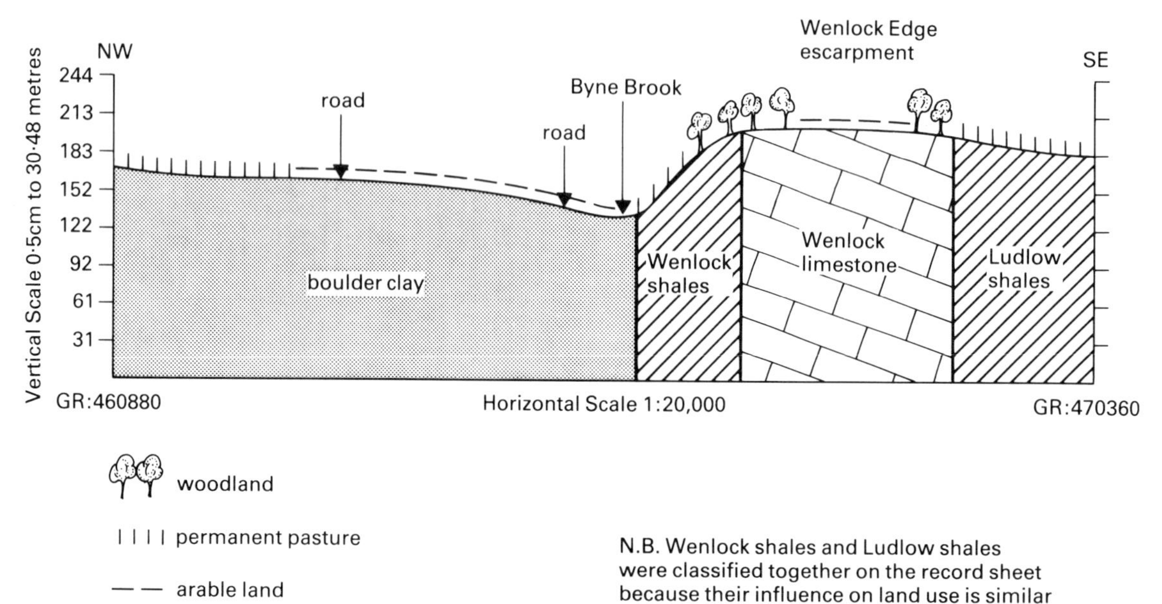

A useful method of summarising your results is to draw a neat, accurate and fully labelled cross-section through the area, showing land use, altitude, aspect and geology.

Analysis

You must look carefully at all your evidence and begin by describing your results. It will probably be best to start with a general description of the land use based on the cross-section, and then to discuss each variable in a little more detail using your graphs and diagrams. Your analysis will be of very little value indeed if you simply describe your results and you must go on to consider the reasons why altitude, aspect and geology are important to the farmer. You must also look at the inter-relationships between the variables, for example, the geology may be closely linked to altitude, or aspect or both. Remember to consider the results you actually obtained even though they may not be what you expected. If your arable land is all facing north then look for reasons for this, do not pretend that it faces the sun just because you think it should.

Your results should provide evidence of the influence of physical factors on the land use pattern of the area under study, and from this you should be able to accept or reject your hypothesis.

Remember that a farmer's decision will depend on many factors other than the altitude, aspect and underlying geology of his land (see fig. 1.1); other physical factors such as slope angle or shelter may be important in addition to socio-economic factors. Do not ignore these influences just because you have no statistical evidence for them. Finally, you should summarise your findings and consider whether you can accept or reject your initial hypothesis.

Suggestions for further study

The influence of economic factors may be incorporated into a study of this kind when data are being recorded at the sample points. Such factors might be distance to the farmhouse (measured in km or time) or distance to the nearest road. However, if distance to the farmhouse is recorded care must be taken to find out to whom the land belongs. It does not always go with the nearest farmhouse, and a visit to all the local farmers is therefore essential before starting the work. This problem could best be solved by selecting an area belonging to one landowner only.

Further reading

R Dougherty, *Analysis of Land Use Data*, No 27 in Geographical Association Occasional Papers 1976

P McCullagh, *Data Use and Interpretation*, OUP 1974 (chi-square test pp. 6–12)

M Mowforth, *Statistics for Geographers*, Harrap 1979 (chi-square test p. 30)

P Toyne and P Newby, *Techniques in Human Geography*, Macmillan 1971 (pp. 78 and 82 for proportional circles)

M J Walker, *Agricultural Location*, Basil Blackwell 1977

2 River variables

Suitable for group or individual study

River channels are part of a self-regulating system. They alter their characteristics or variables to accommodate changes in *input* (discharge). When discharge (Q) remains constant (over a period of time or along a stretch of stream between two tributaries) the river's width (w), depth (d), cross-sectional area (A), wetted perimeter (p) and velocity (v) can all be observed to vary in relation to each other. An important equation showing the relationship between some of these variables is:

$Q = A \times v$

where

$A =$ width $\times$ mean depth ($\bar{d}$)

(mean depth, $\bar{d}$, has to be calculated from all your depth readings)

2.1 River variables

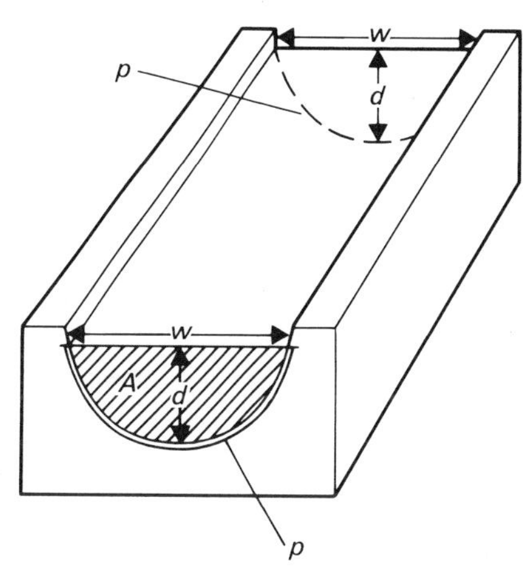

Possible hypotheses

When discharge (Q) is constant:

1 Width (w) varies inversely with mean depth ($\bar{d}$).
2 Cross-sectional area (A) varies inversely with velocity (v).
3 Velocity (v) increases when slope increases.
4 Wetted perimeter (p) varies inversely with velocity (v).

2.2 Results collection

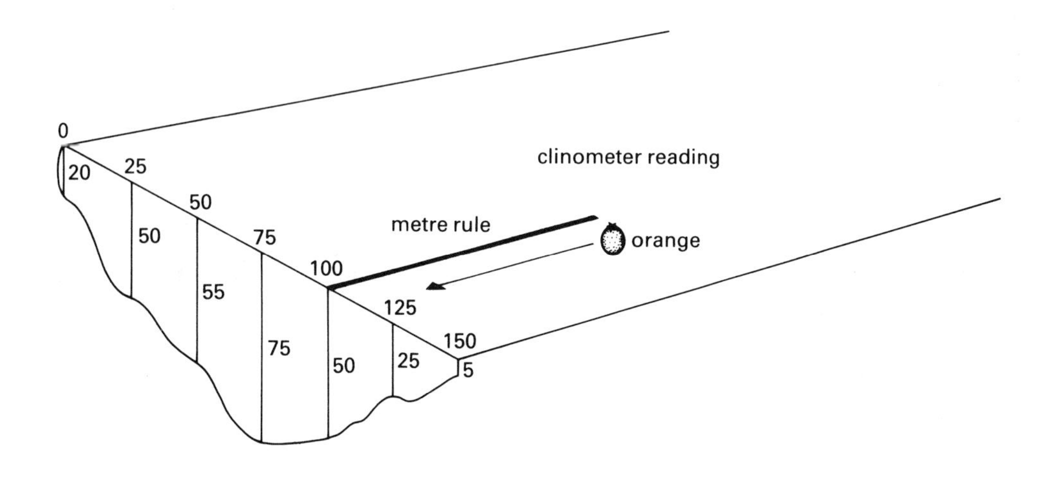

Results sheet

Site	Width (m)	Depth (cm)				Velocity (secs/m)				Slope (°)		
1	1·5	20	50	55	75	20						
		50	25	5								

Location

These variables can be measured only in a stream not deeper than wellington boots or fisherman's waders. The chosen section must be between two tributaries in an attempt to keep discharge constant. The stream should preferably show obvious channel variations, for example riffles and pools. The river should not be sampled at times of very high or low discharge as this will lead to difficulties in data collection.

Data collection

First assemble the equipment:
prepared results sheet (fig. 2.2)
metric tape
metric rule or ranging rod
clinometer
weighting corks, or an orange, for measuring surface velocity (students may wish to devise their own current meter for measuring internal velocities)
stop-watch
surveyor's chain (optional)

Select ten sites along the stream, using a suitable method of sampling (see Appendix A).

At each site in turn:

1 Stretch the tape across the stream from right to left bank. Record the distance in metres.
2 Beginning at the right bank, and starting at 0 on the tape, take a depth measurement in cm every 25 cm across the tape. Record the depth at the left bank even though this may be less than 25 cm from the previous depth reading.
3 Record the velocity at each 25 cm interval along the tape. Place the metre rule on the water surface at right angles to the tape and upstream of it. Float the cork (or orange) over 1 metre next to the rule and note the velocity in seconds per metre. You will need to convert this later into *metres per second*.
4 Record the slope at each 25 cm interval over a distance of 1 metre using the clinometer. Do this at the same time as the velocity reading before proceeding along the tape.
5 If you have a surveyor's chain, allow it to rest on the wetted perimeter (p). If you have no chain, p can be measured on the cross-sectional diagram (see below).

2.3 Calculated results sheet

Site	Width w (m)	Mean depth $\bar{d}$ (m)	Cross-sectional area A (m^2)	Wetted perimeter p (m)	Mean velocity (m/s)	Mean slope (°)
1						
2						
3						
etc.						

Processing data

Prepare a calculated results sheet (fig. 2.3).

1 Calculate the mean depth for each site ($\bar{d}$). Convert from cm to metres.
2 Calculate the cross-sectional area (A) for each site where $A = w \times \bar{d}$.
3 Calculate the mean velocity for each site. Remember that the measurement was in seconds per metre. Convert your results into metres per second using the formula velocity = distance ÷ time *before* you calculate the mean. For example, Site 1, reading 1:
distance = 1 metre
time = 20 seconds
$v = \frac{1}{20} = 0.05$ m/s
4 Calculate the mean slope.
5 If you have no surveyor's chain the wetted perimeter may be calculated from a cross-sectional diagram (see below). Convert to metres.
6 Draw a cross-sectional diagram for each site.
(*a*) Draw a line scale on 2 mm graph paper (1 cm = 25 cm).
(*b*) Draw channel width.
(*c*) Mark off width at 1 cm intervals. Draw a vertical line using the same scale to represent the depth at each interval.
(*d*) Join together all depths.
(*e*) Using a sheet of paper or a length of cotton measure the wetted perimeter.

2.4 Cross-sectional diagram for site 1

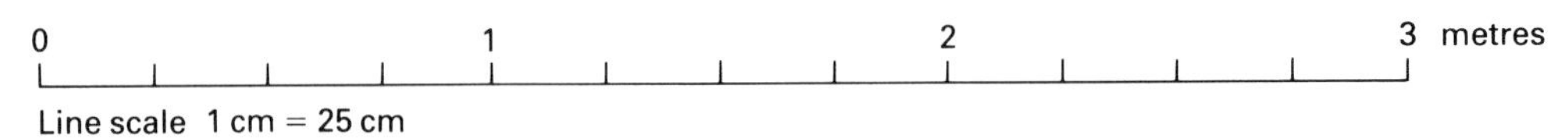

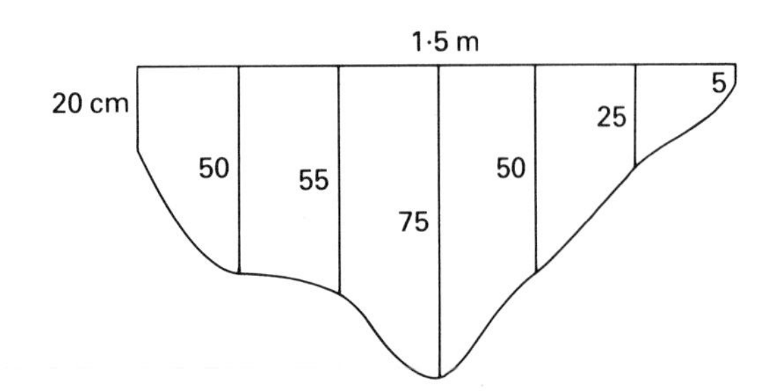

Presentation of results

Plot a scatter graph (2.5) to test each of the four hypotheses proposed. Test each hypothesis with Spearman's rank correlation as shown in the worked example.

2.5 Scatter graph to test the hypothesis 'when *Q* is constant, *A* varies inversely with *v*'

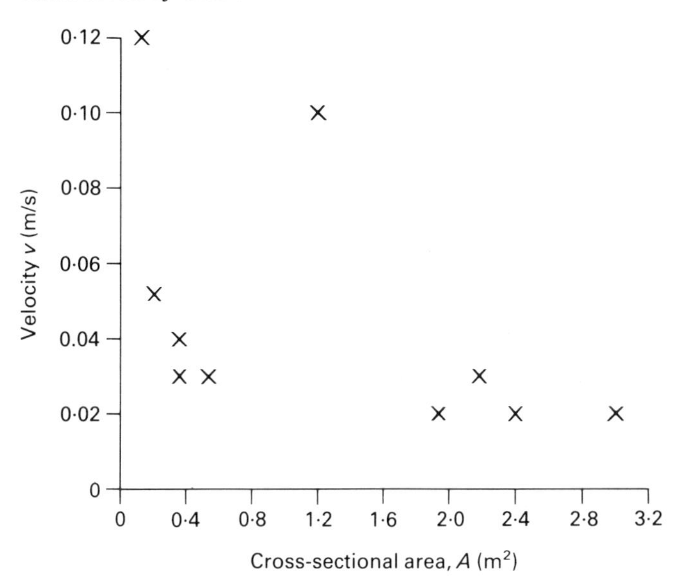

Spearman's rank correlation test

This is used to measure the degree of correlation between two variables (see Toyne and Newby, *Techniques in Human Geography*).

First you must set a hypothesis: 'When discharge is constant cross-sectional area varies inversely with velocity.'

Then prepare a table showing the pairs of variables and give a rank to the variables in each column, 1 for the largest, 2 for the next etc.

Site	Area (m^2)	Rank	Velocity (m/sec)	Rank	d	d^2
1	1.20	5	0.10	2	3	9
2	1.96	4	0.02	9	−5	25
3	2.90	1	0.02	9	−8	64
4	2.18	3	0.03	6	−3	9
5	2.4	2	0.02	9	−7	49
6	0.35	7	0.03	6	1	1
7	0.53	6	0.03	6	0	0
8	0.34	8	0.04	4	4	16
9	0.18	9	0.05	3	6	36
10	0.12	10	0.12	1	9	81
						$\Sigma d^2 = 290$

Substitute Σd^2 in the following formula

$$r_s = 1 - \frac{6\Sigma d^2}{n(n^2-1)} \quad \text{(where } n \text{ is the number of pairs)}$$

$$= 1 - \frac{1740}{990} = -0.75$$

The r_s value must now be looked up in the significance tables (see Appendix C). When $n = 10$ a value of -0.75 is significant at the 0.01 level. This means that we can be certain that in 99% of cases the correlation did not occur by chance. The *negative* value of r_s (-0.75) indicates a very good inverse correlation, hence we can accept our original hypothesis.

Analysis

Take each hypothesis in turn. Comment on its validity in terms of your scatter graph and/or correlation test. Then accept or reject your hypothesis. Try to account for any rejection you have had to make; consider any problems you encountered in collecting and recording your data.

You may find that, contrary to your expectations, there is no correlation between slope angle and velocity. You would then need to ascertain what other factors affect velocity when Q is constant. One of these is the *roughness factor*. Further reading might suggest how it could be measured.

Suggestions for further study

1 The mean velocity and standard deviation could be calculated for each site. Histograms could be plotted showing the distribution of velocities about the mean.

The relationship between channel cross-section and velocity could be investigated – what type of channel cross-section gives greatest/least deviation from the mean velocity?

2 The ideal channel shape for water transport, giving lowest energy losses through friction is

$w = 2d$

Investigate the relationship between channel shape and velocity at each of your ten sites.

Further reading

F Clegg, *Simple Statistics*, Cambridge University Press 1982

R Kay Gresswell, *Physical Geography*, 2nd edn, Longman 1979

M Morisawa, *Streams: their Dynamics and Morphology*, McGraw Hill 1968

P Toyne and P T Newby, *Techniques in Human Geography*, Macmillan 1971 (gives a useful explanation of Spearman's rank correlation test)

3 Consumer behaviour: an examination of Huff's probability model

Suitable for group or individual study

Topic for study

People's shopping habits are an interesting area for study and consumer behaviour can easily be studied through fieldwork. Many shoppers live in or close to small or medium-sized towns and therefore need to visit towns of a 'higher order' from time to time.

D L Huff's probability model (1962) attempts to predict the probability of people purchasing goods in different shopping centres. The model takes into account the relative attraction of different shopping centres in an area and the distance that would have to be travelled in each case. Stated in its simplest form the formula is:

$$\text{Probability of visiting town A} = \frac{\dfrac{\text{Attraction of town A}}{\text{Distance to town A}}}{\dfrac{\text{Total attraction of all towns under study}}{\text{Total distance to all towns under study}}}$$

Huff assumes that shoppers consider both distance and the attraction of towns, and that less 'attractive' towns and those at greater distances will be visited less frequently than larger towns or those which are comparatively close. The attraction of the town may be measured in terms of population size, number of shops, turnover in £,000 per year or any other suitable index. Similarly distance may be measured not only in miles or kilometres but also in terms of time or cost.

Possible hypothesis

Huff's probability model can be used to predict consumer behaviour.

Location

Clearly the choice of settlements to study is important and requires careful thought. As a rough guide you should choose a small or medium-sized town with a population of between 5,000 and 15,000. The town chosen must have a number of larger towns around it and within a radius of 25 miles. It is important that the town chosen is large enough to have plenty of shoppers and yet should provide only a limited range of shops so that people need to visit other towns for higher order goods. Beware of tourist centres and early closing days!

Data collection

Two sets of data are required. Firstly data which can be used in Huff's formula to calculate the *theoretical* movement of shoppers and secondly data to find the *actual* preferences of shoppers. Thus the validity of Huff's probability model can be tested.

Data for use in Huff's formula

To calculate the theoretical movement of consumers various measures of attraction and distance are needed. Sources of this information are:

(*a*) The *Census of Distribution* found in most Reference Libraries. This contains statistics such as number of shops, numbers employed in shops and turnover in £,000s per year

(*b*) The *AA Members' Handbook*. A useful source of population and mileage statistics for individual towns.

Find at least two measures of attraction and two of distance for each of your selected towns.

Data concerning the actual preferences of shoppers

Collect answers to the following questionnaire from 200 or more shoppers. This questionnaire was used in Bakewell, Derbyshire and of course you will need to substitute the names of your towns.

1 Do you live in or close to Bakewell? (If 'No', discontinue the interview.)

2 When shopping outside Bakewell which of these towns do you usually visit? (Show a list of your selected towns on a card, about 15 cm × 20 cm, covered in clear sticky plastic (fig. 3.1).

3.1

Ashbourne
Chesterfield
Matlock
Buxton
Sheffield

3 Why do you choose this town in preference to the others? (Mark the preference sheet, see fig. 3.2, to show the interviewee's reasons, but *do not suggest* any of the reasons. Usually you will be able to record more than one reason for each interviewee.)

When you are conducting your survey you should try to obtain an unbiased sample. For example, car owners will

3.2 Preference sheet (question 3 of questionnaire)

Reason		Ashbourne	Chesterfield	Matlock	Buxton	Sheffield
1	Shops with quality and fashionable goods		𝍸 𝍸 ||||	𝍸 |	||||	𝍸 𝍸 𝍸 |||
2	Shops offering competitive prices and bargains		𝍸 ||	||||	𝍸 ||||	𝍸 ||
3	Reliable shops		𝍸 ||			𝍸 𝍸
4	Large variety and choice of shops		𝍸 ||			𝍸 ||
5	Easy access and/or parking		𝍸 𝍸 |||	𝍸 |||		𝍸 |
6	Ability to combine shopping with other activities	||	𝍸 ||	𝍸 |||	𝍸 ||||	𝍸 𝍸 |
7	Atmosphere and appearance of town	||			𝍸 𝍸	
8	Other reasons					|
	Total (177)	4	55	26	32	60

3.3 Attraction and distance table for towns near Bakewell

	Attraction		Distance from Bakewell	
	Number of shops*	Turnover in £000s*	in miles	bus time (min.)
Bakewell	65	1,663	—	—
Ashbourne	98	2,448	19	159
Chesterfield	736	28,998	13	40
Matlock	214	4,599	8	21
Buxton	257	6,015	12	25
Sheffield	4,817	160,118	16	50

*From 1971 Census of Distribution

3.4 Theoretical probability of visiting towns*

	Raw probability	Percentage probability
Ashbourne	0.0567	1.26
Chesterfield	0.6222	13.77
Matlock	0.2940	6.51
Buxton	0.2354	5.21
Sheffield	3.3089	73.25
Total	4.5172	100%

* using number of shops and milage from Bakewell.

probably have different shopping habits from people who rely on public transport. It would be best to collect a systematic sample, questioning for example, every fifth passer-by, but in reality this can be difficult to operate. However, you must try to interview a cross-section of people including different age groups and both sexes. Don't stand by a car park or at the bus station.

Processing data

1 Tabulate your attraction and distance measures as in table 3.3. Select one measure of attraction and one measure of distance from your data and use these in Huff's formula to find out the theoretical consumer movement. For example, using the number of shops and mileage from Bakewell:

$$\text{Probability of visiting Ashbourne} = \frac{98 \div 19}{6187 \div 68} = 0.0567$$

Then convert the raw probability to percentage probability:

$$\text{Percentage probability of visiting town A} = \frac{\text{Raw probability for town A}}{\Sigma \text{ Raw probability for all towns}} \times 100$$

2 Repeat this procedure using several different measures of attraction and distance. Present all your results in a table. You may decide to choose one set of these theoretical results to compare with the actual pattern, or to average your findings.

3.5 Results for Bakewell, 1981 (sample of 200)

	Theoretical consumer movement*		Actual consumer movement	
	%	rank	%	rank
Ashbourne	1.26	5	1	5
Chesterfield	13.77	2	39	2
Matlock	6.51	3	10	3
Buxton	5.21	4	9	4
Sheffield	73.25	1	41	1

*using number of shops and mileage from Bakewell. This information could also be shown in a desire line diagram, as in fig. 3.6.

Presentation of results

1 Analyse the responses to question 2 of the questionnaire, and work out the *actual* pattern of consumer behaviour. Tabulate these results together with the theoretical consumer movement which you have just calculated. Rank your results (see table 3.5).

2 Analyse the responses to question 3 of the questionnaire (reasons for the choice of alternative shopping centres), to find out what shoppers actually think are the attractions of the towns being studied. This can be compared with the measures you used to calculate Huff's probability values. Tabulate the results of this part of the questionnaire.

3.6 Flow diagrams to compare theoretical and actual consumer movement

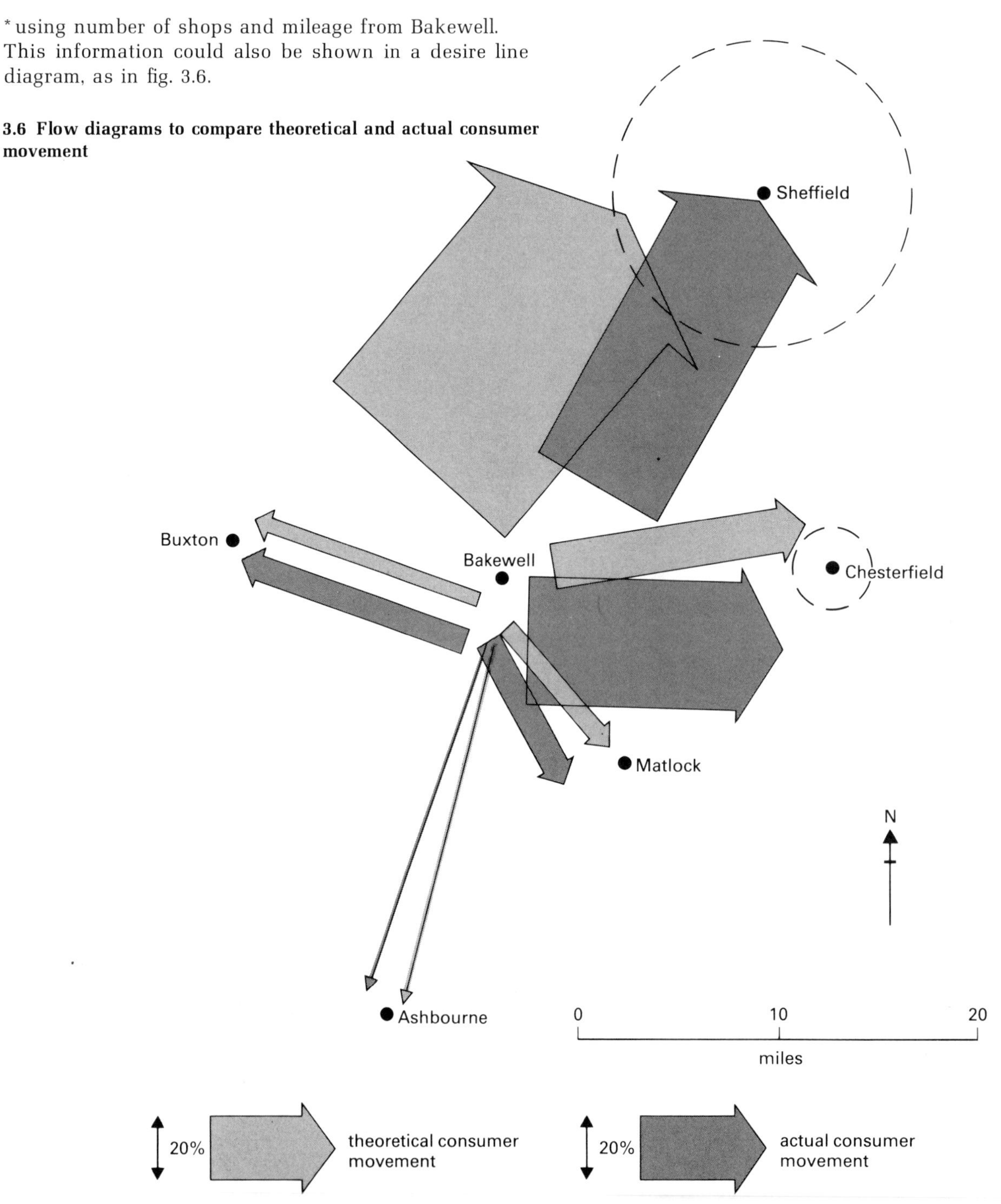

3.7 Divided column diagram showing reasons for consumer movement

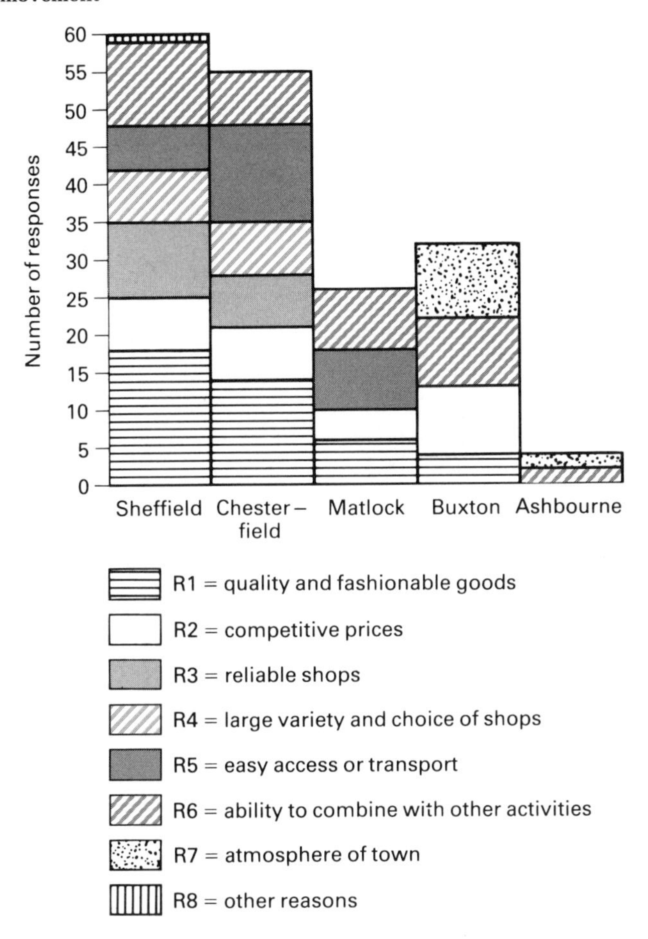

3 Present this information in a divided column diagram (see fig. 3.7). Remember that you will have recorded several reasons for each person interviewed, so total figures will not be the same as the total number of people interviewed.

Analysis

Your results may or may not support your original hypothesis. In the example given, Sheffield was found to be much less popular than predicted although the rank order of popularity was the same.

First describe your results carefully, then you must try to explain the actual pattern which you have observed. Use the information collected about shoppers' reasons for choosing one town rather than another in your explanation. Huff's basic assumption was that consumers responded to distance and attraction; however, in order to measure attraction for fieldwork purposes, only economic factors have been taken into account. Your fieldwork might suggest that other factors, perhaps social factors, are more important to shoppers. Remember to consider your hypothesis in your conclusion.

3.8 Perception index questionnaire

	Ashbourne	Chesterfield	Matlock	Buxton	Sheffield
Access (bus, car)	1	5			
Shops	2	5			
Character of town	4	4			
Appearance of town	4	3			
Total	11	17			

Suggestions for further study

1 Your results might suggest that the measures of attraction selected to work out predicted consumer movement were not appropriate. You could devise an index of your own which incorporates the perception people have of the towns you are studying, such as fig. 3.8. Score 1 to 5 in each category for each town using a questionnaire. At least 30 people should be questioned and their scores should be totalled to provide a perception index. This index can then be used in Huff's formula.

2 A similar study could be carried out using one type of shop only, for example jewellers. The *Census of Distribution* contains statistics concerning individual shops. It might be interesting to compare the usefulness of Huff's Model using a selection of different shops instead of a whole town.

Further reading

B J L Berry, *Geography of Market Centres and Retail Distribution*, Prentice Hall 1967

J A Everson and B P Fitzgerald, *Settlement Patterns*, Longman 1969

D L Huff, 'Defining and estimating a Trading Area' in section 4 of *Analytical Human Geography* edited by P J Ambrose, Longman 1969

W V Tidswell, *Pattern and Process in Human Geography*, University Tutorial Press 1976

4 Upland ecosystems

Suitable for group or individual study

Topic for study

The aim is to investigate the relationships in an upland ecosystem. One definition of an ecosystem is as follows: 'The ecosystem . . . seeks to explain the inter-relationships between all the different members of a community and the relationship with all other physical factors in the environment.'
Geographers are probably more concerned with examining the relationship between plant communities and their environment than in looking at relationships between species in one community. A community is a group of plants co-existing in a specific locality. Some of the factors leading to the establishment of different communities are:

climate
soil type and drainage
topography (altitude and slope)
aspect
geology
grazing animals

A study of an ecosystem should attempt to measure all the above variables. In practice it is usually necessary to consult the work of others – for instance it is not practical or relevant to collect climatic data over a few days as long-term data will be more significant in influencing the development of a plant community.

Possible hypothesis

Variations in slope, aspect, soil characteristics and the influence of grazing animals are responsible for the co-existence of many different plant communities in a small area of (the selected location).

N.B. The effect of grazing animals may be quite considerable in determining what species are present. Sheep for example prefer certain species and graze selectively within a community. Those species disliked by the sheep have an unfair advantage over the preferred species and may be dominant in the community. Those species disliked by sheep may begin to invade neighbouring communities, thereby changing the composition of these communities.

Location

In order to make the study manageable choose a small upland area, not more than 80 square metres, which contains several different and readily recognisable plant communities, e.g. bracken, ling, rush, mat grass. Be sure to obtain permission to carry out your survey on private land. You should endeavour to choose a site in which some of the variables are known to be constant and need not be measured, such as geology and climate. These two variables however are important in determining the range of species present in the whole area. Climatic and geological data can be obtained beforehand from libraries.

Data collection

Equipment

clinometer – slope angle
compass – aspect
auger – soil sample
metre rule – soil depth/humus depth
small, ready labelled plastic bags – soil collection
quadrat – composition of vegetation community
plastic markers – to mark out sample points

Preliminary method

1 Consult OS maps, geology maps and climate statistics to select several possible locations. Visit each of them and choose one that offers good scope for testing your hypothesis.
2 Using a plant identification book identify what you consider to be the 15 to 20 commonest species.
3 Prepare a data sheet similar to fig. 4.1. (The species included are those found on the Silurian slate area of the Lake District.)
4 Prepare a data sheet (fig. 4.2) for recording measured variables (keep climate and geology constant).

4.1 Plant recording sheet

Site	Polytrichum	Sphagnum	Soft rush	Cotton sedge	Ling	Heather	Mat grass	Fescue	Meadow grass	Other grasses	Bracken	Bilberry	Tormentil	Heath bedstraw	Other species	Total no. of species
1																
2																
3																
etc.																

4.2 Measured variables

Site	Position on slope (top etc.)	Aspect and angle	Soil depth (cm)	Depth of organic layer (cm)	Soil pH	Soil moisture	No. of sheep faeces
1							
etc.							

5 Decide how many sample points you need to provide sufficient data for analysis. If you intend to use correlation techniques such as Spearman's rank you should have no more than 30 points. If you intend to use the chi-square test you should not have less than 20 points.

6 Choose a sampling procedure (see Appendix A). As you will have chosen an area with several different plant communities a stratified sampling method may be best (see Appendix A).

Field method

Mark out the sample points. At each point or site:

1 Make a *quadrat survey*. Throw the quadrat. Decide whether you intend to investigate *local shoot frequency*, which records hits to leaves and shows *cover*, or *local rooted frequency*, which records hits to individual plants and shows *density*.

When the quadrat lands record the leaf (or plant) found at each intersection on your plant recording sheet (25 'hits', see fig. 4.3).

Complete the 'total number of species' column on your sheet.

4.3 Quadrat

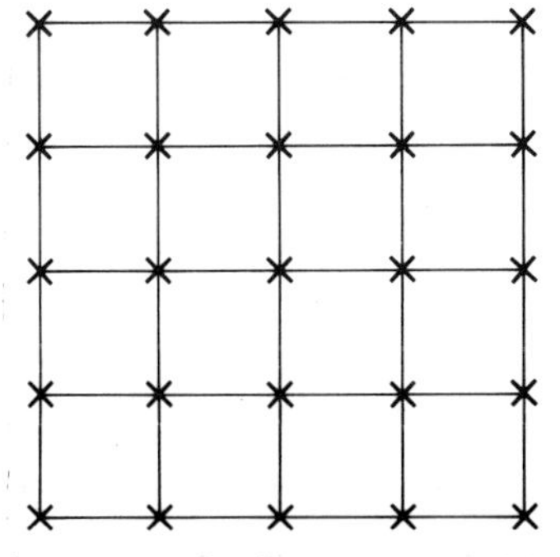

2 Count the number of sheep faeces under the quadrat and measure the aspect and angle of slope and record these.

3 Use the auger to collect a soil sample from the centre of the quadrat. Record the total depth and the depth of the organic layer (humus at the surface which has not been incorporated). Take a sample of soil from the A horizon, which lies beneath the organic layer. Do not touch it with your fingers or you will affect the acidity readings.

4 Indicate the position of the site on the slope.

Processing the data

Laboratory work

First test part of each soil sample for pH. (See Hanwell and Newson, *Techniques in Physical Geography*). Then find the moisture content as a percentage of the weight using the rest of each sample as follows. (Ask the Science department if you can use their oven and scales. The Laboratory Technician will advise you about temperatures.)

1 Weigh a tin or evaporating dish.

2 Place soil in container and weigh. Label container to show which site it represents.

3 Place labelled containers in oven and leave for at least 24 hours at a low temperature.

4 Weigh container of dry soil.

5 Substitute your weights in the following formula.

$$\text{percentage moisture content of soil} = \frac{W-D}{W-T} \times 100$$

where W = weight of tin and wet soil
D = weight of tin and dry soil
T = weight of tin

Refining of data

1 Calculate the mean number of plants in each species in each plant community you have identified.

2 Work out the mean slope position and aspect for each plant community.

3 Calculate the mean pH, moisture, number of faeces, soil depth and depth of organic layer, slope angle.

4 Prepare and complete a summary sheet which shows both mean conditions and the range of conditions in each community (fig. 4.5).

4.4 Example of the means for ling communities

Site	Polytrichum	Sphagnum	Soft rush	Cotton sedge	Ling	Heather	Mat grass	Fescue	Meadow grass	Other grasses	Bracken	Bilberry	Tormentil	Heath bedstraw	Other species	Total no. of species
1	3				15	6						1				4
2	1				12	8	2					2				5
3	4				10	6	2					2	1			6
4	5				14	6										3
Mean	3.25				12.75	6.5	1					1.25	0.25			4.5

4.5 Example of a summary sheet showing results for bracken communities

Slope position (mean)	Aspect (mean)	Slope angle		pH		Soil moisture	
		mean	range	mean	range	mean	range
Middle	SE	13°	10°–17°	5.7	5–6	17%	10–27%

Note that you can use both raw and refined data in your analysis.

4.6 Divided column graph showing relationship between plant communities and slope angles

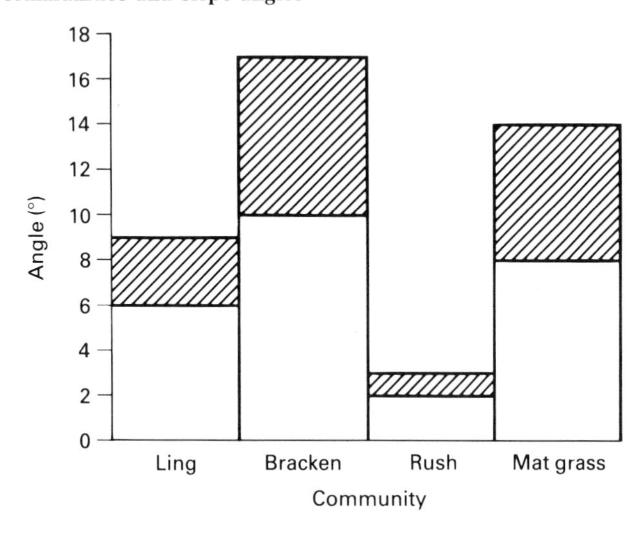

Presentation of results

1 Mean slope angles can be represented on a block graph. Range of angles data can be plotted as shown in fig. 4.6.
2 Draw similar graphs to show number of faeces present.
3 Draw scatter graphs or use Spearman's rank correlation to examine some of the relationships between your variables. Here are some possibilities. Use raw data for these tests.
   (*a*) The lower the pH value the lower the number of species present.
   (*b*) The lower the pH value the greater the depth of the organic layer.
   (*c*) The steeper the slope the drier the soil.
4 Star diagrams can be used to show the mean composition of each community. Draw spokes only for those species that are present, otherwise you will have problems when constructing the polygons. Choose a suitable scale for your data. Keep it constant on all the diagrams.
5 Dispersal diagrams can be used to show the relationship between non-quantified and quantified data (see fig. 14.8). e.g. plant community / soil acidity
   slope position / slope moisture content
6 Note that it would be valid to examine the relationship between any of the variables you have collected. Remember however that you are attempting to examine the *reasons* for co-existence of different communities in a small area.

4.7 A star diagram for ling communities

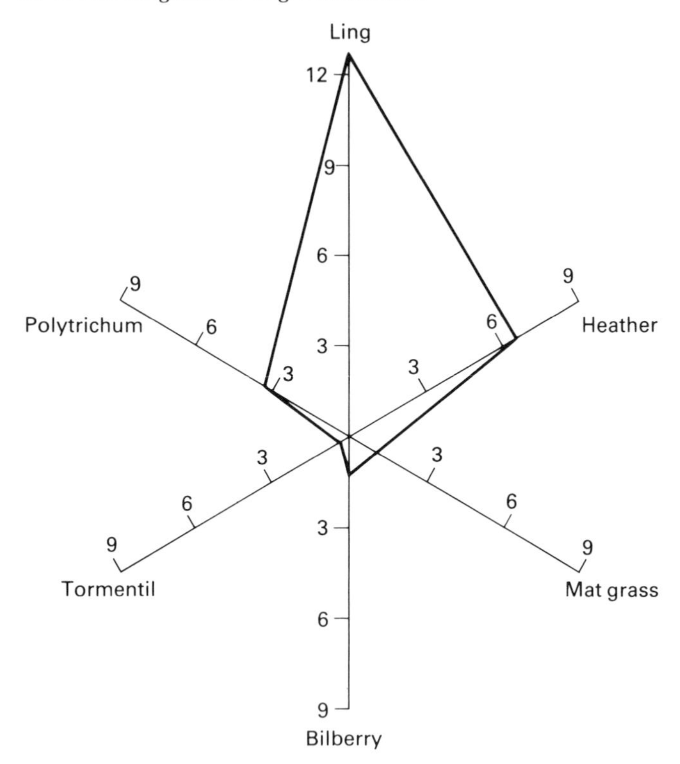

Analysis

With the aid of the frequency polygons and the summary chart *describe the differences* in the composition of each community and the conditions in which it thrives.

Refer to your dispersal diagrams, graphs and Spearman's rank correlation tests and comment on those relationships *which appear to be significant* in controlling the distribution of different communities.

Accept or reject your original hypothesis. Should you be forced to reject it, attempt to put forward a new hypothesis from your data.

Suggestions for further study

1 Consult standard texts such as Eyre's *Vegetation and Soils* and Pearsall's *Mountains and Moorlands*. Compare the conditions you have encountered with those ascribed to such communities in textbooks.
2 Investigate the ways in which individual species are adapted to the conditions in which they grow. Find out how they alter the conditions and pave the way for plant succession.
3 Make a comparison between shoot frequency and rooted frequency methods of reading a quadrat (see Hanwell and Newson, *Techniques in Physical Geography*).

Further reading

S R Eyre, *Vegetation and Soils*, Edward Arnold 1975
J D Hanwell and M D Newson, *Techniques in Physical Geography*, Macmillan 1973
W H Pearsall, *Mountains and Moorlands*, Collins 1972

5 The Central Business District

Suitable for group or individual study

Topic for study

The Central Business District (CBD) of a town is usually a distinct area exhibiting the following characteristics.

1 *Accessibility*: Evidence of this includes large numbers of vehicles and pedestrians and a large proportion of land being used as roads and pavements.
2 *High rise buildings*: Competition for land and high rents has led to developments of multi-storey buildings.
3 *Commercial land use*: Most buildings are used for certain types of function, particularly shops and offices.
4 *Lack of open space.*
5 *Lack of industry*: Exceptions include small residual industries and small warehouses, and those industries depending on a central location such as newspaper production.
6 *Lack of housing*: Exceptions include residual housing and in some large centres penthouse flats.

Using these characteristics it is possible to measure a number of criteria in an attempt to delimit the CBD of a town (fig. 5.1). In addition some measures will enable identification of the inner and outer core, the boundary zone and the transition zone.

It may not be possible to identify all these zones in all towns, and some zones may exist only in some parts of the CBD. It may also be possible to identify 'quarters' where particular functions cluster together e.g. professional offices.

Possible hypotheses

1 The CBD can be delimited using the following criteria: land use, building height, rateable value, pedestrian flow, and parking restrictions.
2 It is possible to distinguish different zones within the CBD.

Location

This exercise works well in a small town where the CBD is compact or based on a single main street. In a large town or city where a survey of the whole CBD would be very time consuming you could study a number of transects from the centre to the edge of the CBD. You should choose main roads radiating out from a central point, such as the Town Hall, in different directions.

Data collection

It is important to choose at least four different criteria to measure and these should relate to different characteristics of the CBD.

1 Prepare a base map of the area you have decided to study. This could be based on the 1:2,500 OS map, or it could be a simple sketch map of your own. You will need several copies of this map.
2 Record the land use of all the buildings on your base map. Make sure you record land use on all floors of the buildings because ground floor use is often different to use of higher floors. The main land uses will be commercial (shops and offices) although some other functions may be observed. Remember to extend your survey to the surrounding areas of industry, housing and transition to enable you to determine the edge of the CBD. Devise a simple classification for recording land use, for example

   shop — industry
   pub, restaurant, or cafe — housing
   office — vacant building
   bank or building society — derelict land

   An alternative method of recording this information using graph paper is shown in fig. 5.2.
3 Record building height by counting the number of storeys. Use the same base map for this data or record it on graph paper as shown in fig. 5.2.
   (a) Draw a column 1 cm wide down the centre of a sheet of graph paper. This represents the road.

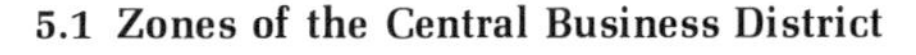
5.1 Zones of the Central Business District

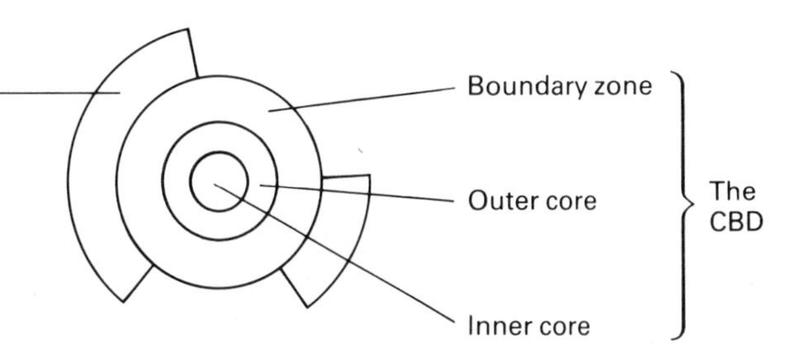

5.2 Recording height and land use on graph paper

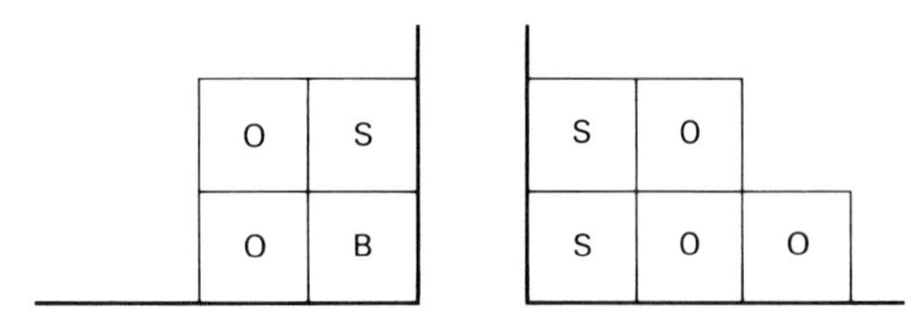

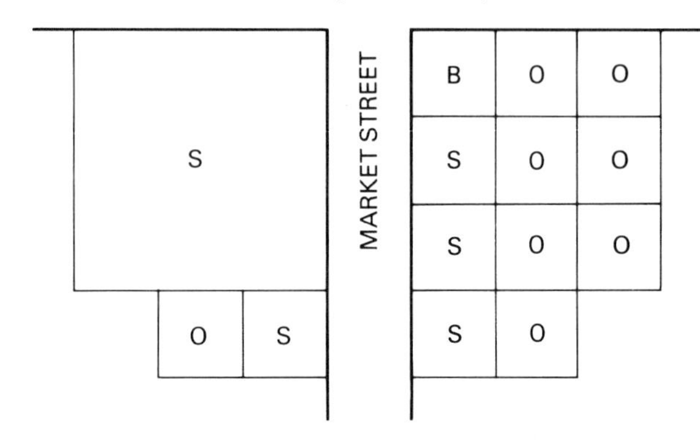

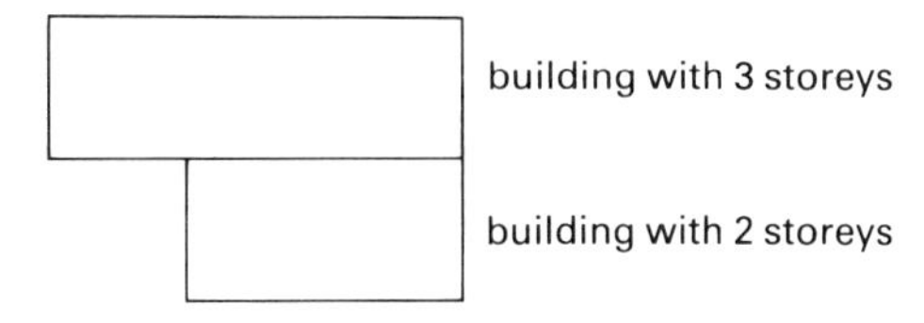

(*b*) Use a vertical scale of 1 cm = 1 building except for large stores which have a large footage e.g. department stores, when you may use several cm/building.

(*c*) Use a horizontal scale of 1 cm = 1 storey.

(*d*) Put in road junctions as they occur, using the scale 1 cm = 1 road junction. Do not attempt to make opposite sides of the road 'match'.

(*e*) Land use and building heights on roads entering at right angles must be plotted on a separate sheet.

4 Find and record rateable values for a sample of buildings in your study area. First select at least twenty buildings and record the address of each one, e.g. 16 High Street. Visit the local Rates Office which will probably be in the town hall and find the *gross rateable value* of each of your sample of buildings. These values cannot be compared with each other unless some account is taken of the size of the premises. A simple method is to measure the front footage of each building in order to establish a *rate index*.

$$\text{Rate index (in £ per metre)} = \frac{\text{Rateable value (in £)}}{\text{Front footage (in metres)}}$$

Other methods of comparing rateable values are discussed in Lowry, *World City Growth*, pp. 41–7.

5 Record pedestrian movement which provides a measure of accessibility. Choose a number of locations throughout the study area and count pedestrians passing each point for a specified time period, such as five minutes. Select locations away from road junctions and away from bus stops and record the numbers of pedestrians passing in both directions. This exercise is best done with a group of people who can count at different locations simultaneously. It can be done by one person if the exercise is done on the same day and within a limited period of time. The survey must not spread over both quiet and busy times or the results will not be comparable with each other.

6 Record parking restrictions on a base map, using symbols to represent double or single yellow lines, parking metres, urban clearways etc.

Processing data and presenting results

If this exercise has been done by a group the information must be exchanged between members of the group. The results can be presented using a series of maps and diagrams.

1 *Present a land use map* showing the ground floor land use (fig. 5.3). Colour this, thinking carefully of the colours you use. These should show a contrast between functions typical of the CBD and those which would not usually be found in the CBD. Thus commercial land use could be shown in shades of red, orange and yellow, with industry and housing in contrasting colours such as blue and green. Building heights and parking restrictions can be shown on the map.

2 *Draw an urban transect.* Where the fieldwork includes information collected along a transect or down one main street, information can be presented on graph paper, see fig. 5.4. This shows building height, drawn to scale, and land use on all floors. Use the same colours as your land use map. In addition, pedestrian flow can be shown using arrows where the width is drawn to scale.

3 *Draw an isometric diagram* which represents information in three dimensions, on isometric graph paper (fig. 5.5). This is an effective way of showing building height and land use on all floors. Use the same colours to show land use. In addition either pedestrian flow or parking restrictions could be incorporated into the map.

4 *Map the rate index* Record this onto either your urban transect or your isometric diagram. If there is a gradual change in the rate index you may be able to construct an isopleth map.

5 *Draw a map or a tracing overlay showing the boundary of the CBD* based on all your results. If possible, show other zones which can be distinguished such as the inner and outer core, the boundary zone and the transition zone. Use all your information to help you to decide where the boundaries occur. In the diagrams shown the zones have been marked on figs. 5.3 and 5.4.

Analysis

Used in combination the sets of data collected in the field will probably enable you to delimit the CBD and to distinguish zones within it. Each set of data collected should

5.3 Transect through Sheffield CBD showing ground floor land use, building height (in storeys) and parking restrictions

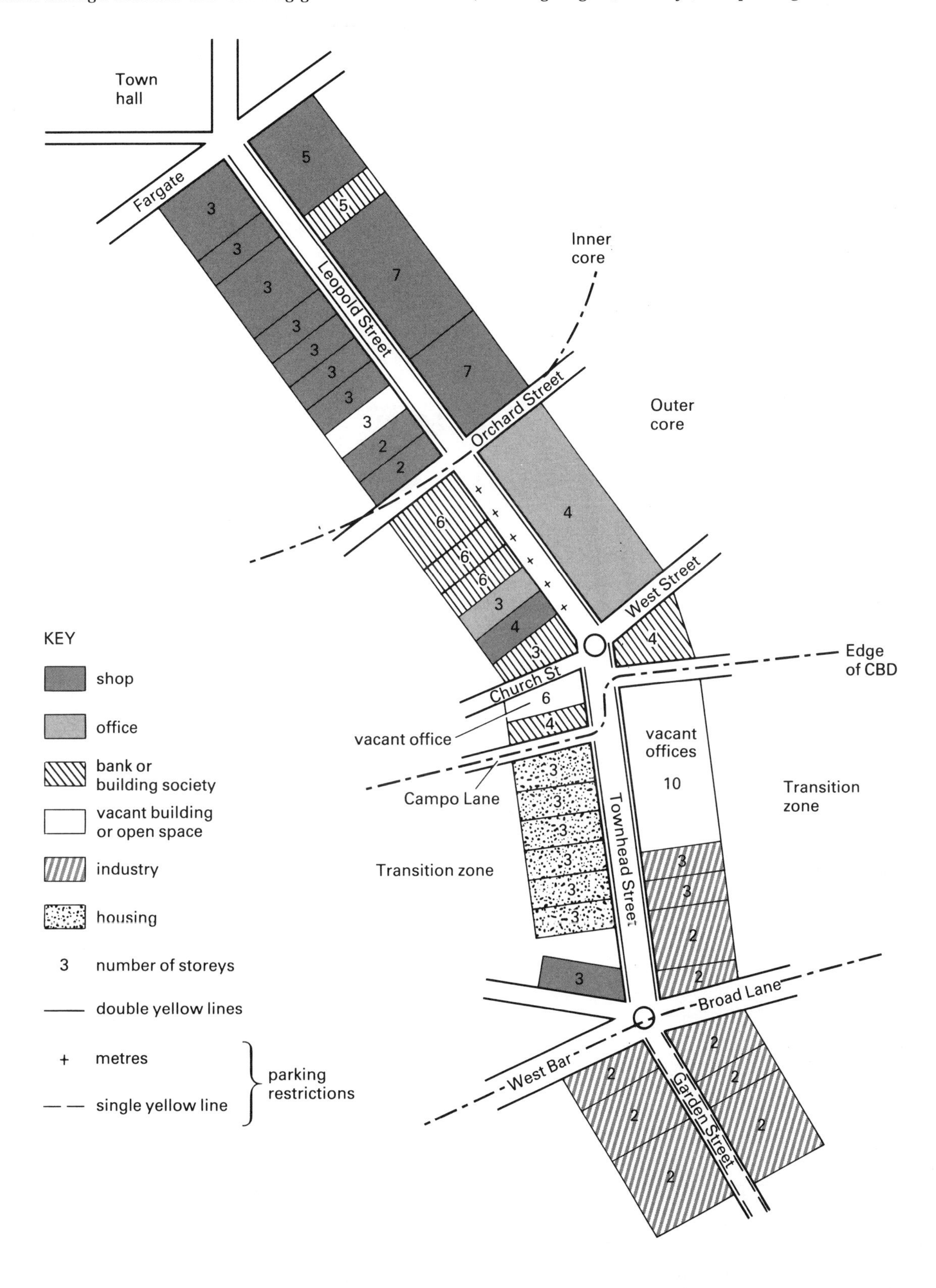

5.4 Urban transect showing land use, building height, pedestrian flow and rate index in Sheffield CBD

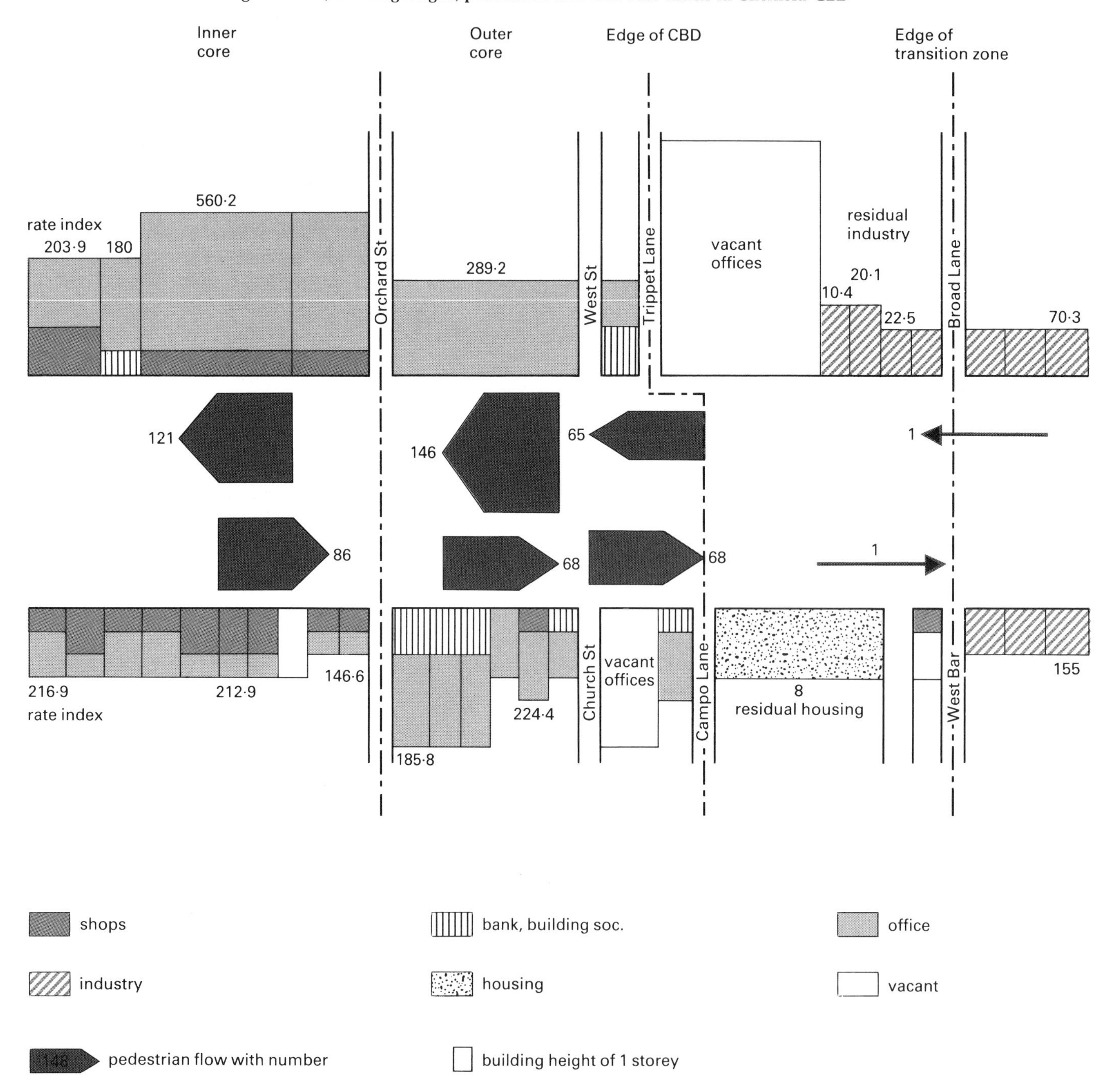

be discussed separately using the maps and diagrams you have drawn. Thus you should attempt to describe and explain the patterns of land use, building heights, pedestrian flow and parking restrictions.

In some cases there will have been problems concerning data collection which should be discussed. In other cases you may find that the pattern you expected has not been found. For example, building heights may increase towards the edge of the CBD if this area has had recent new buildings. Similarly the rateable value and thus the rate index may reflect the age of the buildings. Discuss points such as these fully.

Having discussed the data sets separately you must then look at the results overall. You should describe where you think the edge of the CBD lies and explain carefully why you have chosen this position. Similarly you must describe the extent of any other zones you have identified and justify your decisions in each case. Any unusual features or anomalies should be noted and explained, for example, residual housing or industry.

Suggestions for further study

1 Attempt to correlate groups of data collected in the field

5.5 **Isometric diagram showing land use and building height for part of Sheffield CBD**

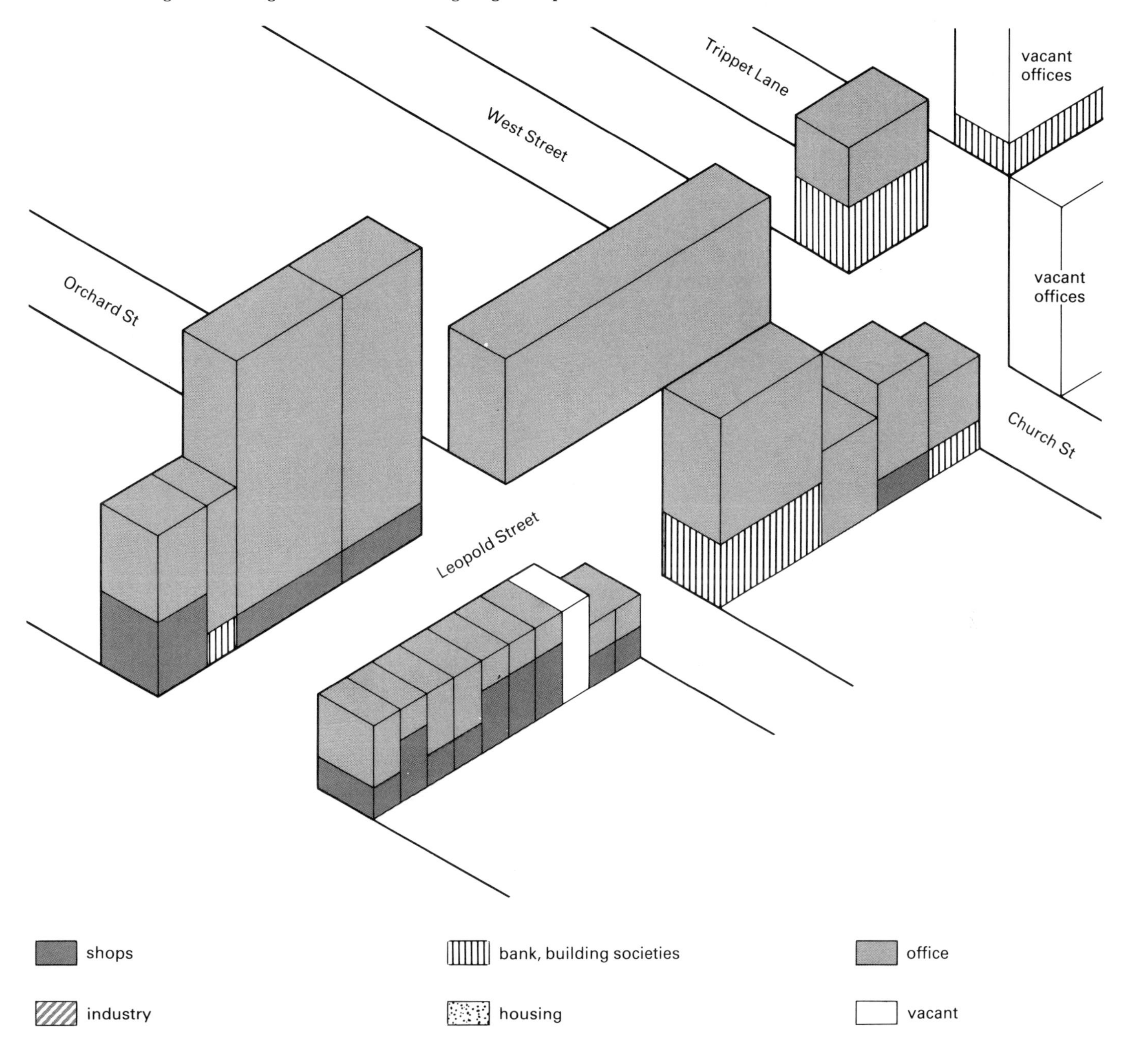

and thus incorporate some statistical analysis into your work. For example, compare

(*a*) building height and distance from central point;

(*b*) pedestrian flow and distance from central point.

First draw scatter graphs to show the information and if a correlation is shown then test using Spearman's rank correlation (r_s) (see chapter 2).

2 Compare the CBDs of two towns or cities which are different in size. You could test a hypothesis such as 'different zones within a CBD can be identified in large towns but not in small towns'.

3 Functional zoning within a CBD could be studied, for example looking at the occurrence of chain stores or lawyers' offices. The nearest neighbour analysis can be used to look at distributions. Toyne and Newby, *Techniques in Human Geography*, p. 151, suggest a modified version of this technique using 'reflexive pairs'.

Further reading

K Briggs, *Fieldwork in Urban Geography*, Oliver & Boyd 1971

P Daniel and M Hopkinson, *The Geography of Settlement*, Oliver & Boyd 1979

J H Lowry, *World City Growth*, Edward Arnold 1975

P Toyne and P Newby, *Techniques in Human Geography*, Macmillan 1971

6 Plant colonisation and succession on sand dunes

Suitable for a pair of students or a group

Topic for study

Newly exposed or developed land surfaces soon undergo invasion by plants. Those plants which first gain a foothold on the new surface, known as initial colonisers, are usually able to adapt to severe or hostile environmental conditions. As they grow they alter these conditions by the addition of humus; their root systems bring increased stability; the shade cast by leaves and stems cuts down evaporation and increases the moisture content of the material in which they grow.

Eventually other more demanding plants are able to invade and displace the initial colonisers; they also change the environment and are replaced in turn by another community. When the vegetation has become stable, over a long period of time and is not being invaded by newcomers, it is said to have reached 'climatic climax'.

Although the study of vegetation change through time is difficult, it is possible to establish the plant succession by examining *spatial* changes. Coastal sand dunes usually consist of dunes of differing ages, and a study of plant distribution on the dunes will therefore show vegetation change through space and time.

Possible hypotheses

1 The nearer a sand dune is to the sea, the more likely it is to support a simple plant community which has not significantly changed its environment.
2 The nearer the sea, the fewer species are present in the community.

Devise five hypotheses yourself which suggest likely relationships between

1 Moisture content/percentage vegetation cover
2 Salt content/distance from the sea
3 Number of species present/salt content
4 Number of perennials/distance from the sea
5 Distance from sea/acidity

Location

Any accessible coastal sand dunes which have not undergone afforestation.

Data collection

Equipment

metric tape (30 m minimum length)
compass
clinometer
quadrat
trowel
labelled plastic bags for samples
plant identification sheet

Some of the commonest species found on sand dunes are noted in table 6.1. You would find it helpful to draw and describe each plant to make your own identification sheet.

6.1 Common sand-dune species

Plant	Life cycle
Sea rocket	Annual
Mouse eared chickweed	Annual
White grass	Annual
Ragwort	Perennial
Marram grass	Perennial
Sea spurge	Perennial
Creeping fescue	Perennial
Rest harrow	Perennial
Hawthorn	Perennial
Dwarf willow	Perennial
Lichen	

Choice of sites

You will need to complete several transects (4–6) from foreshore to fixed dunes. Fixed dunes are almost completely vegetated and show no signs of new sand having accumulated. The distance between the transects will be determined by the width of the foreshore. Your transects should be regularly spaced.

6.2 Arrangement of transects

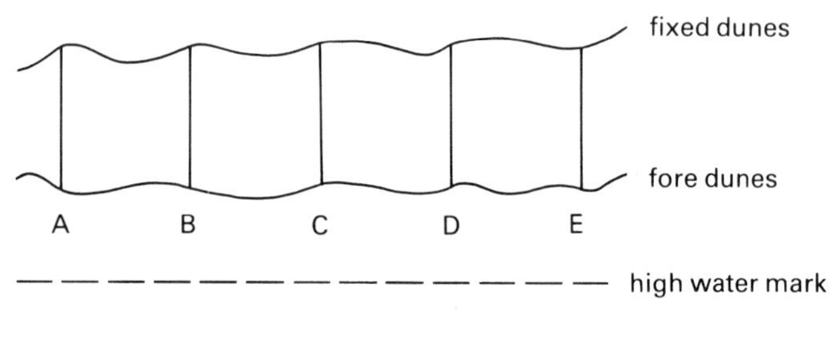

6.3 Sample transect

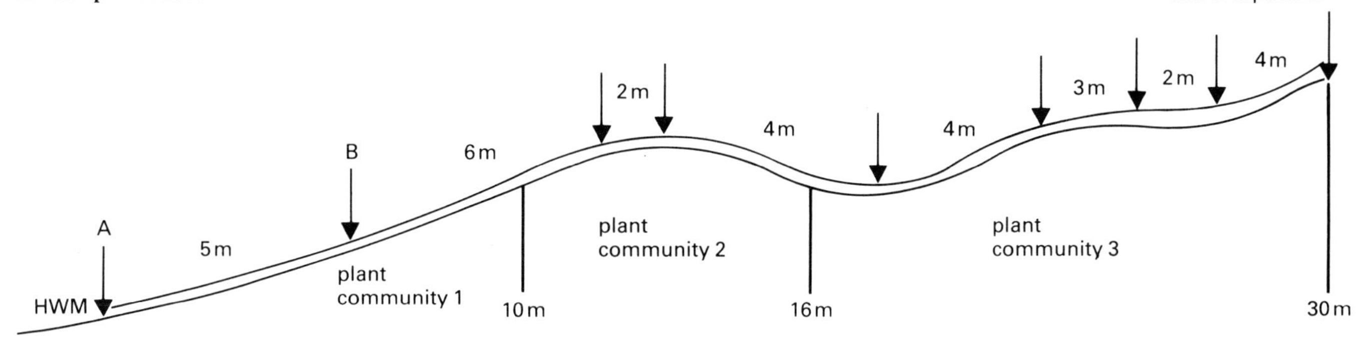

6.4 Sample slope recording sheet
Transect no 1

Reading	Foresight (°)	Backsight (°)	Average angle (°)	Distance (m)	Rising/ Falling	Change in community
1	10	12	11	5	+	
2	13	14	13.5	6	+	10 metres
3	2	3	2.5	2	+	
4	5	5	5	4	−	16 metres
5	9	11	10	4	+	
6	2	2	2	3	+	
7	2	1	1.5	2	−	
8	7	6	6.5	4	+	30 metres

Method

1 Take a compass reading to establish direction. Keep checking this reading as it is easy to move off course.
2 Locate the HWM and begin at right angles to it.
3 Stretch the tape to its fullest extent up the dune from the HWM.
4 At each break of slope (indicated ↓) take a foresight and backsight reading using the clinometer. Person A stands at HWM and sights onto the eyes of person B (who should be the same height as A). This is the *foresight*. Person B then reads back to person A (backsight). Calculate the average angle. Record these values (see table 6.4).
5 The distance between A and B, i.e. the length of the slope facet, is recorded as shown.
6 Record whether the slope is rising (+) or falling (−)
7 *At the same time*, each time a new species of plant appears, or a species disappears, record the distance *from the start* at which the change was noticed.
8 Complete a vegetation recording sheet (fig. 6.6). Throw the quadrat back along the tape into the community which has been left (fig. 6.5). (If your quadrat has 25 intersections multiply by 4 to get % results.)
9 Collect a sample of sand from beneath the quadrat. You will need sufficient to test separately for pH, moisture and salt.

Processing data

Divide each sand sample into three. Find the percentage moisture content of each sample 1 (see chapter 4).

Find the percentage salt content of each sample 2 as follows.

1 Weigh container.
2 Place sand in container and leave to dry.
3 Weigh sand sample in container.
4 Moisten sand thoroughly with tap water and shake well.
5 Pour through filter paper to leave sand as residue (salt should have filtered out in water).
6 Leave sand to dry and remove from filter, putting it back into its original container.
7 Reweigh dried sand.
8 Substitute your weights in the following equation:

$$\text{percentage salt content} = \frac{S-W}{S-T} \times 100$$

where S = weight of container and sand sample
W = weight of container and washed sand
T = weight of container

Test each sample 3 for pH (see Hanwell and Newson, *Techniques in Physical Geography*).

Presentation of results

1 Stick together several large sheets of graph paper so that you can draw out all your slope transects above one another on one sheet. Place the transects 12 cm apart to allow for graph drawing. Do not attempt to draw the transects on separate sheets or comparisons will be difficult.
2 Using a protractor, draw out the slope profile for each transect. Choose a suitable scale after consulting your data for all transects. For example, if the total length of

6.5 Sampling vegetation

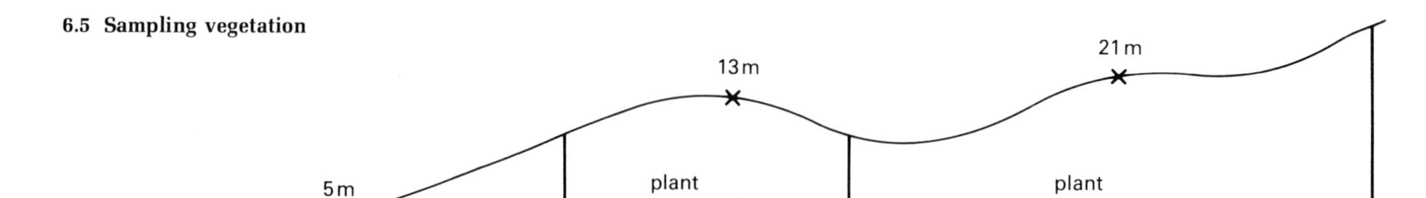

6.6 Vegetation recording sheet
Transect no 1

Distance from start	Sand	Sea rocket	Chickweed	White grass	Ragwort	Marram	Sea spurge	Rest harrow	Hawthorn	Dwarf willow	Lichen	% Cover	Laboratory tests % Moisture	% Salt	pH
5 m	60	20	15			5						40	5	45	7
13 m	30	10	5		5	50						70	8	43	7
21 m	5	0	5	10		70	10					95	15	36	6.7

longest transect is 225 m and the smallest slope facet is 2 m, a suitable scale would be 2 mm = 1 m, while if the total length of the longest transect is 30 m and the shortest slope facet is 2 m a suitable scale would be 5 mm = 1 m.

(*a*) Put a dot (1) at the left-hand side of your graph paper, 6 cm from the top.

(*b*) Place the protractor horizontally with the centre over the dot. Note that reading (1) is *rising*.

(*c*) Measure mean angle (1), which is 11°, and draw a line at that angle to represent distance (1), which is 5 m.

(*d*) Move the protractor to the end of the distance line (2). Repeat with mean angle (2), which is 13.5°, and distance (2), and so on.

(*e*) Delete all angle numbers and distance numbers from the completed transect.

(*f*) Mark the quadrat reading sites (X).

3 Mark off all the changes in community as shown on figs. 6.3 and 6.5 on each of your drawn transects.

4 At the points where your quadrat readings were taken draw a divided column below your transect. Choose a suitable scale for the width and depth of each bar. Species should always be plotted in the order of their occurrence with distance from the sea, omitting species as necessary. Use a colour code on your column to indicate perennials, annuals and biennials.

5 Draw two small pie charts above each transect at the point where the quadrat readings were taken to show percentage moisture and salt. You may need to overlap your pie charts if space is tight.

6 Write the pH reading at each point.

7 Use scatter graphs or Spearman's rank correlation to test hypothesis 2 and the hypotheses you have devised for yourself.

8 The results of your scatter graphs and correlation will probably have produced some negative and some positive correlations. You might find it helpful to annotate a *systems diagram* (fig. 6.8) by showing positive and negative *feedback*. The annotation in fig. 6.8 has been started for you. It shows that an *increase* in distance from the sea correlates with a *decrease* in salt content.

Analysis

Analyse the columns on each individual transect to see whether there is a plant succession (disappearance of certain species, appearance of others). Compare the transects to see whether there is a similar succession. If it exists you should be able to describe the stages in the succession on the dunes and also provide the answer to the first part of hypothesis 1.

Environmental changes through a succession will be shown on the transect diagram and will have been examined in the scatter graphs, correlation tests and systems diagram. You should therefore be able to write a conclusion to the second part of hypothesis 1.

The divided bar graphs will also yield information for hypothesis 2.

Suggestions for further study

1 Investigate the relationship between individual species and specific environmental conditions. For instance, does

6.7 Presentation of results

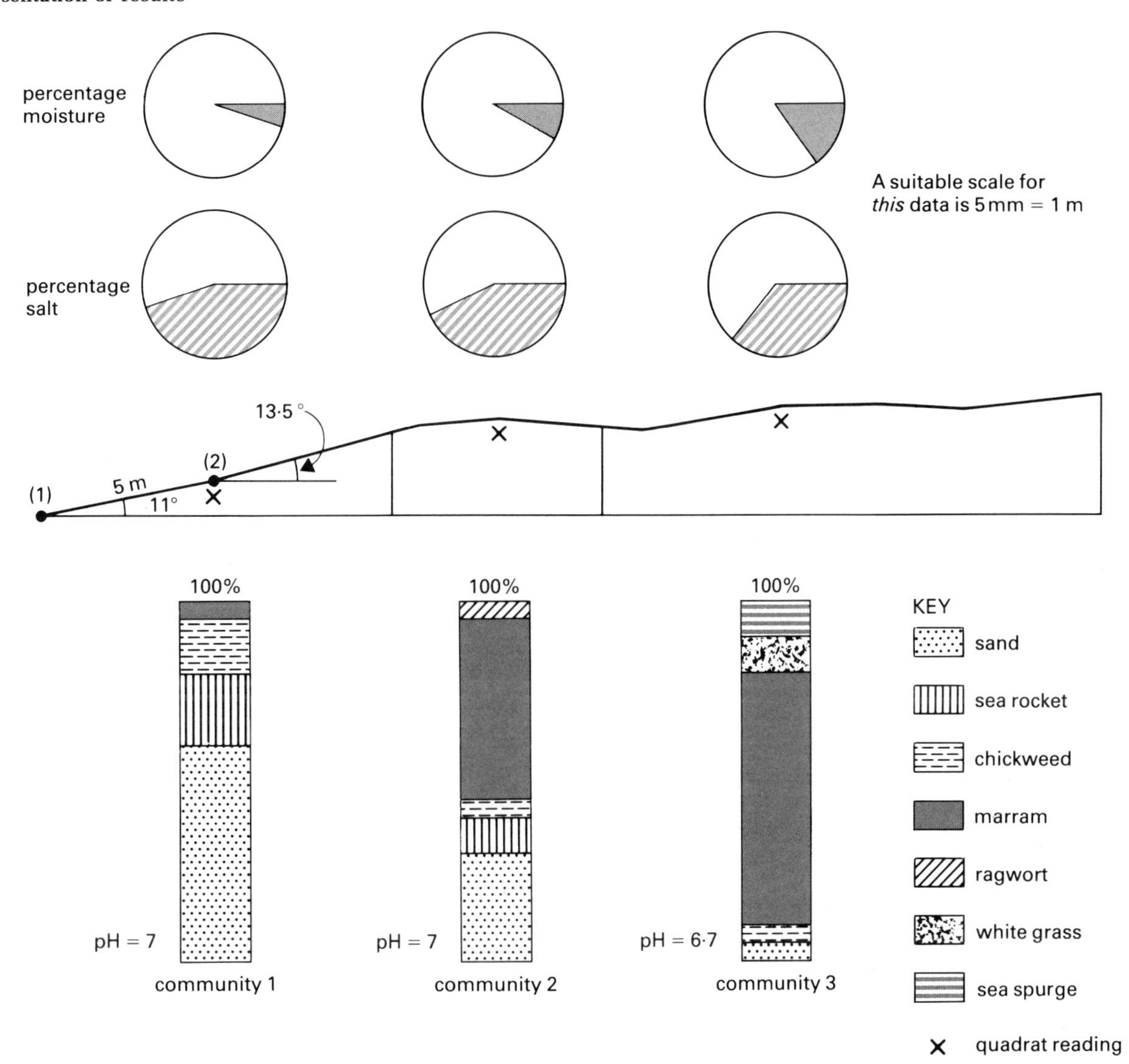

marram die out when salt content falls and percentage moisture increases?

2 The data you have collected for slope angle could be used to investigate changes in dune shape with distance from the sea. You could find out whether fixed dunes have steeper or gentler slopes than foredunes. Is there a correlation between slope angle and percentage plant cover?

3 Use a set of sieves (see chapter 8) to investigate shell content/100 g sand. Is there a correlation between shell content and pH? If so, how do plant communities adjust to changes in the shell content of the dune?

6.8 Annotated systems diagram

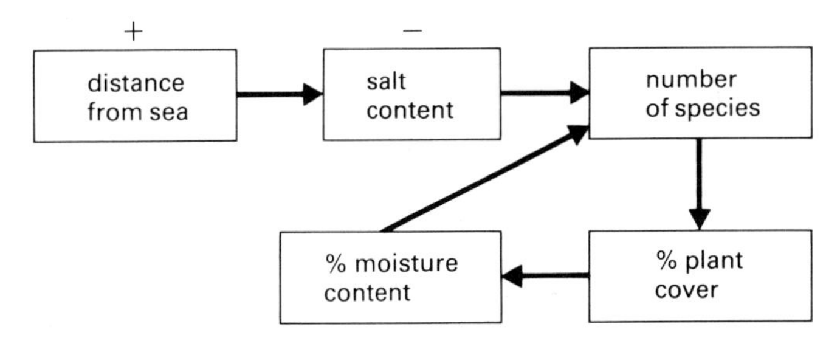

Further reading

S R Eyre, *Vegetation and Soils*, Edward Arnold 1975

J D Hanwell and M D Newson, *Techniques in Physical Geography*, Macmillan 1973

B J Knapp, *Practical Foundations of Physical Geography*, Allen & Unwin 1981

J A Steers, *The Sea Coast*, Collins

7 Shopping centres within a town

Suitable for individual or group study

Topic for study

All towns contain shopping centres of different sizes; in addition to the Central Business District there will be a number of other shopping centres ranging from large suburban centres to corner shops. These centres will be used more frequently than the CBD and for lower order goods, although some specialist shops or even chain stores may be found in the larger suburban shopping centres.

A hierarchy of shopping centres of different sizes can be identified in most towns. In large towns such as London, Birmingham or Greater Manchester, as many as six levels in the hierarchy may exist but smaller towns will have fewer levels.

First order — *Isolated retail outlets* or corner shops. These are scattered throughout residential areas.

Second order — *Retail clusters*. Small groups of shops selling food and household goods.

Third order — *Neighbourhood shopping centres*. A larger number of shops selling a large range of convenience goods. Some services such as banks and dry cleaners will also be available.

Fourth order — *Major suburban centres*. These provide a wide range of goods and services and may also have chain stores.

Fifth order — *Regional shopping centres*. These only exist in very large towns and provide large numbers of shops including major chain stores and department stores. Some may even have pedestrianised areas and shopping precincts.

Sixth order — *The Central Business District*. The central shopping area for the town.

Centres at different levels in the hierarchy have different characteristics. In addition to the variation in the number and types of shops there will also be differences in the habits of shoppers using them. Suburban centres will usually attract people from greater distances than neighbourhood centres and these people will probably use the larger centres less frequently. Retail clusters on the other hand will largely be used by people living nearby who call in almost everyday.

This area of study provides an interesting basis for two pieces of fieldwork which could be done separately or together. The first exercise is to attempt to identify a hierarchy of shopping centres within a town and the second concentrates on looking at differences between shopping centres at different levels of the hierarchy.

1 ESTABLISHING A HIERARCHY OF SHOPPING CENTRES

Possible hypothesis

A hierarchy of shopping centres can be identified within a town.

Location

You must choose a town which has a number of different shopping centres of different sizes in addition to the central shopping area in the CBD. You will need to visit at least 20 shopping centres in order to identify different levels in the hierarchy.

In large towns and cities it will be impossible to visit all the suburban shopping centres and you should therefore study one region of the town only. Choose main arterial roads to divide your region from other parts of the town. Familiarity with the town is an advantage because you will know where to find the small retail clusters that might be missed by someone who does not have a good knowledge of the town.

Data collection

1 List the shopping centres (at least 20) which you are going to include in your study. It is probably better to omit first order shopping centres (corner shops and very small shopping areas with less than five shops) because of the difficulty of collecting the data. You may also decide to exclude the central shopping area, especially if it is a large centre which would take a long time to survey.
2 Select between ten and twenty indicator functions. These

7.1 Examples of indicator functions

Found in all orders	Found in neighbourhood centres and higher orders	Found in major suburban and regional centres	Found in regional centres only
Greengrocer	Building society	Travel agent	Shoe shop
Newsagent	Chemist	Wool shop	Chain store
Butcher	Dry cleaner	Book shop	Department store
Supermarket		Fishmonger	

must include shops or services which are typical of each different level of the hierarchy. For example, retail clusters (lowest order in your study) will usually include a supermarket, a greengrocer and a newsagent, but a chemist and a dry cleaner will probably only be found in neighbourhood centres and higher order centres. Choose about three indicator functions for each level of the hierarchy (see table 7.1 but remember that other indicator functions could be used).

3 Visit each of the selected shopping centres. Record the number of indicator functions at each centre and the total number of shops. Make a table similar to table 7.2 on which to record your data.

7.2 Data recording sheet

Shopping centre	Greengrocer	Newsagent	Butcher	Supermarket	Building Society	Chemist	Dry Cleaner	Travel Agent	Wool Shop	Book Shop	Fishmonger	Shoe Shop	Other shops	Total number of shops
Ranmoor	|	|	||		|	|								16
Nether Green	|	|					|							10
Hangingwater Rd	|	|	|	|		|								17
Hunters Bar etc.	𝍸 |	𝍸 ||	𝍸	𝍸	𝍸 ||	|||	||	||	|	||	||			59

7.3 Hierarchy of urban shopping centres

		Greengrocer	Newsagent	Butcher	Supermarket	Building Soc & Est Agt	Chemist	Dry Cleaner	Travel Agent	Wool Shop	Book Shop	Fishmonger	Shoe Shop	Total no of Shops A	Sum of Centrality Values B	Centrality Index C	Rank D
1	Ranmoor	1	1	2		1	1							16	18.3	34.3	12
2	Nether Green	1	1					1						10	12.2	22.2	18
3	Hangingwater Road	1	1	1	1		1							17	15.6	32.6	13
4	Hunters Bar	6	7	5	5	7	3	2	2	1	2	2		59	201	260	2
5	Sharrow	2	1	3	1			1						32	25	57	9
6	Banner Cross	2	1	2	2	7	3	1	1	1	2	1	1	52	147.3	199.3	5
7	Ecclesall	1	1	1			1							12	12.7	24.7	16
8	Bents Green	1	1	1	2		1		1					9	28.5	37.5	11
9	Parkhead	1	1											8	4.5	12.5	20
10	Nether Edge	2	1	1	2		1	2		1				17	48.5	65.5	7
11	Broomhill	4	4	1	5	9	2	2	3	1	1	1	2	84	200.4	284.4	1
12	Glossop Road	1	1		1						1			12	19.9	31.9	14
13	Commonside	3	3		1									39	16.4	55.4	10
14	Walkley	7	3	5	2	1	1	1	3	2	1	1	1	65	161.2	226.2	3
15	Upperthorpe	2	2	3	2	1	1							26	31.2	57.2	8
16	Crookes	6	6	9	6	4	1	2		1	1	1		57	142.5	199.5	4
17	Crosspool	2	3	1	2	2	1			1			1	29	63.9	92.9	6
18	Lodge Moor	1	1	1	1		1							7	15.6	22.6	17
19	Fulwood	2	1	2	1	1								14	17.7	31.7	15
20	Upper Fulwood	1	1		1			1						6	15.1	21.1	19
	Total number of shops	47	41	38	35	33	18	13	10	8	8	6	5				
	Centrality value	2.1	2.4	2.6	2.9	3	5.6	7.7	10	12.5	12.5	16.7	20				

7.4 Scatter graph of shopping centres in south-west Sheffield

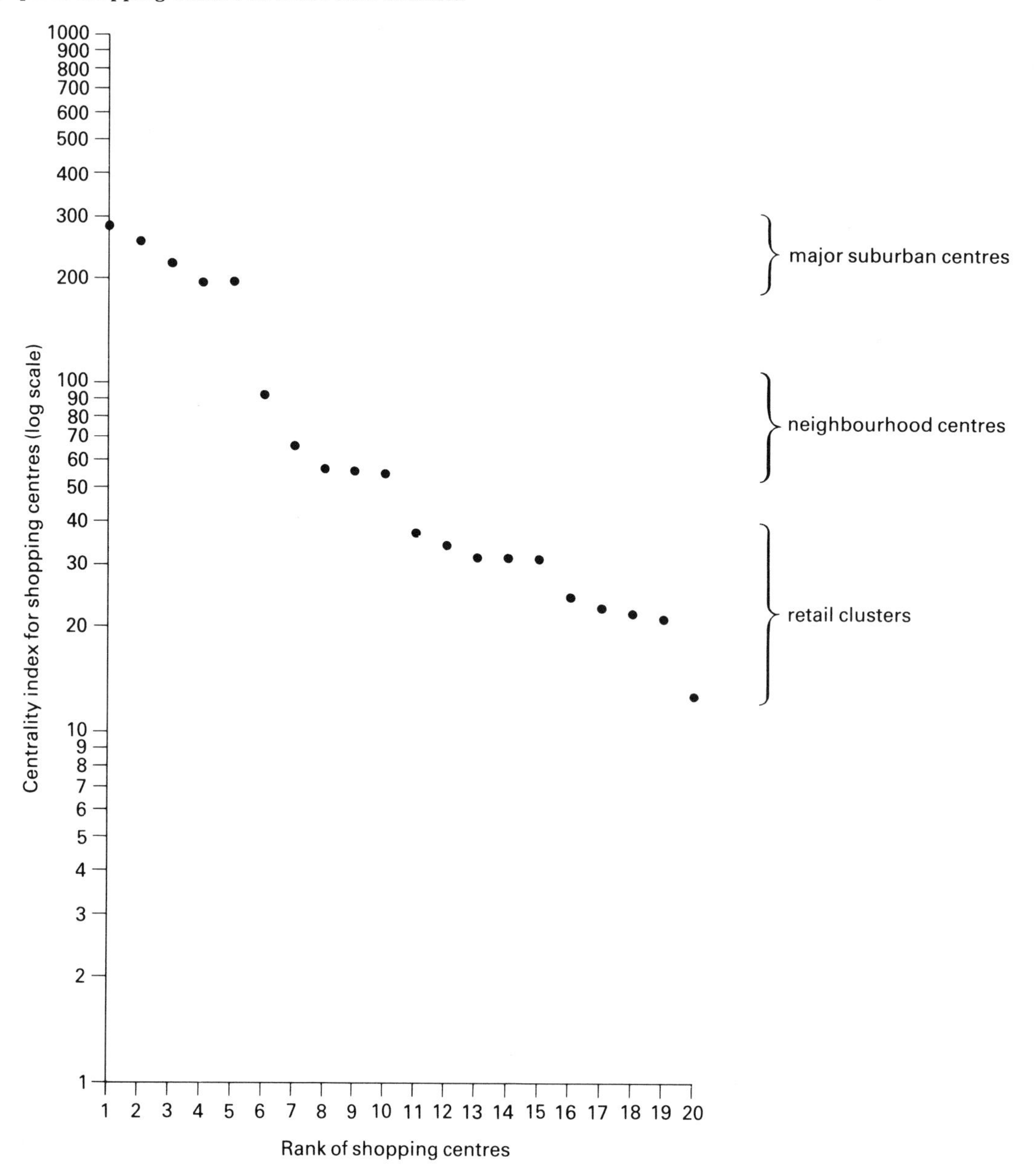

Processing data

Having collected and recorded data about the shopping centres you must attempt to classify them. The number of indicator functions in each centre and the total number of shops should be used to make the classification. The method illustrated here is based on Briggs' idea in his book *Introducing Towns and Cities*. Construct a table similar to table 7.3.

1 Calculate the centrality value for each indicator function and record carefully in the last row.

$$\text{Centrality value} = \frac{100}{\text{Total number of that indicator function found}}$$

So if 41 newsagents are observed, the centrality value for newsagents is $100 \div 41 = 2.4$

2 For each shopping centre score each indicator function observed in that centre using the centrality values. Where there is more than one of the same indicator function in one centre the score will be the appropriate multiple of the centrality value for that function. Thus

1 newsagent scores 2.4,
3 newsagents score 7.2.

3 Add together the scores for the indicator functions for each shopping centre and record (column B). For example, for Nether Green,

Sum $= 2.1 + 2.4 + 7.7 = 12.2$

4 For each centre, add the total number of shops (column A) to the sum of the centrality values (column B). This will give you the centrality index (column C). For Nether Green,

centrality index $= 10 + 12.2 = 22.2$

5 Rank the shopping centres using the centrality index (column D).

7.5 Distribution of shopping centres in south-west Sheffield

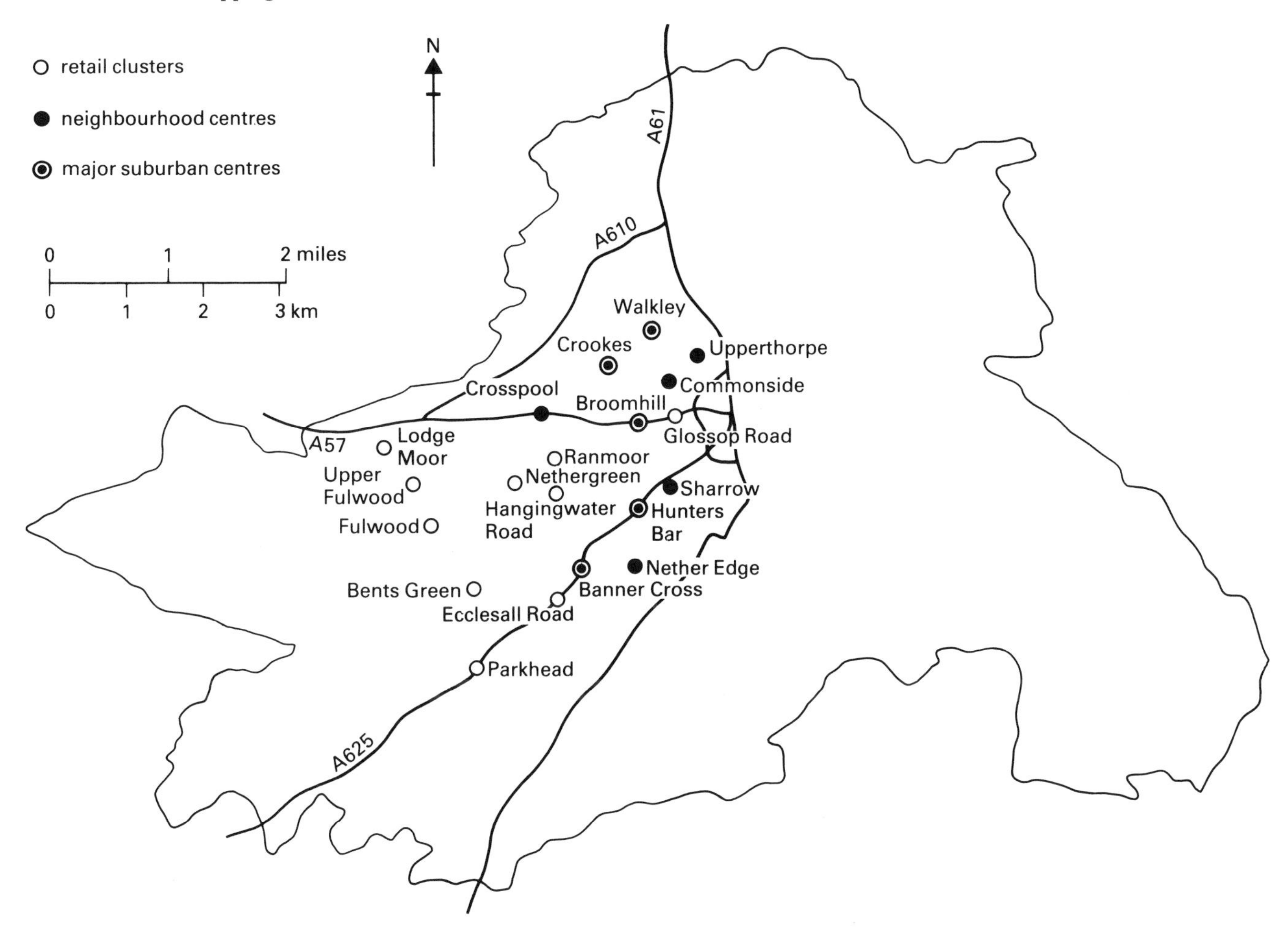

Presentation of results

Your results should be clearly presented in a table and also on log–normal graph paper, as in fig. 7.4. This should enable you to see whether or not a hierarchy of shopping centres exists. In the example given, south-western Sheffield, three levels of the hierarchy can be identified, with five major suburban centres, five neighbourhood centres and ten retail clusters.

The graph shows that the shopping centre ranked 6, Crosspool, is an exception, having a centrality index considerably higher than the others classed as neighbourhood centres. However, table 7.3 shows that it has scored highly because it has a shoe shop although in fact there are only 29 shops in the centre. The small number of shops and the absence of functions such as Travel Agents and Dry Cleaners suggest that it should be classified as a neighbourhood centre and not as a suburban centre.

The division between neighbourhood centres and retail clusters was made between ranks 10 and 11 because the difference in centrality index is 18. No division was made between ranks 15 and 16 because here the difference is only 7.1. One of the disadvantages of a log scale is that it exaggerates differences between smaller numbers; however without a log scale values varying from 12.5 to 284.4 would be difficult to show.

Finally you should draw a map like fig. 7.5 to show the location of the shopping centres you have studied and their level in the intra-urban shopping hierarchy.

Discussion of hypothesis

In discussing your results you should first justify your decisions about which shopping centre is classified at which level in the hierarchy. You may find some centres which do not fit clearly into the hierarchy and you will need to explain how you have classified these. You may be able to do this by reference to your results table or by using your own knowledge about the centre.

It is possible that your results will show a gradual change in size and importance of shopping centres rather than a hierarchy. (In this case there would be no clear 'steps' on your graph.) If this is the case you must try to explain this pattern. It is essential to discuss the results you have obtained and not the result which you would have liked to obtain. Do not imagine a hierarchy if it does not exist.

As a result of your analysis you may decide that some of the indicator functions you chose were not appropriate and you might wish to modify your study by discarding some indicator functions and replacing them by others. The choice of indicator function should be discussed in your conclusion.

When you have mapped the shopping centres and indicated their level in the hierarchy you should comment on this distribution. Again you must describe and explain your results. This may then lead you to suggest a further hypothesis.

Suggestions for further study

This type of study can be used to identify any type of hierarchy, such as:

1 Rural settlements: having established a hierarchy it is interesting to look to see if it corresponds with Christaller's Central Place Theory (see Briggs, *Introducing Towns and Cities*.

2 Parks and recreation areas within towns: facilities such as tennis courts, boating lakes and children's playgrounds can be used as indicator functions (see Bull and Daniel, *The Geography of Outdoor Recreation*).

Further reading

K Briggs, *Introducing Towns and Cities*, Hodder & Stoughton 1974

C J Bull and P A Daniel, *The Geography of Outdoor Recreation*, Occasional Paper 33, Geographical Association Publication 1981

C Whynne Hammond, *Elements of Human Geography*, George Allen & Unwin 1979

2 IDENTIFYING DIFFERENCES BETWEEN SHOPPING CENTRES OF DIFFERENT SIZES

This fieldwork can be done as an extension of the previous study by using the hierarchy which you have established to select suitable shopping centres. However it is not essential to have completed the first study before carrying out this study.

Possible hypothesis

Larger shopping centres have larger trade areas than smaller centres and are used less frequently by customers and for higher order goods.

Location

Choose three or four shopping centres which represent different levels in the hierarchy, for example, a corner shop, a retail cluster, a neighbourhood centre and a major suburban centre. In some towns it may be possible to include the central shopping centre (CBD) but in many towns this will be far too large.

If possible choose shopping centres which serve similar residential areas. The size of the trade area will be affected by the type and density of housing surrounding the centre and the shopping habits of the people who live there.

Data collection

1 Visit each shopping centre. Map the area carefully recording each type of shop. Use a classification such as
  food shops
  clothes and shoe shops
  commercial services (e.g. hairdressers, betting offices)
  professional services (e.g. banks, estate agents)
  cafe, pub, restaurant
  household goods (e.g. hardware or electrical)
  specialist shops (e.g. antiques)

2 Conduct a questionnaire in each of the shopping centres you have selected. At least 50 people in each centre should be questioned and you should choose similar times and days of the week to conduct your questionnaire. Try to interview people of different ages and of both sexes. Some possible questions are given below. Where possible you should record the answers in categories as shown in questions 3, 4 and 5. This is a more accurate means of recording information and makes the later analysis much easier. Devise a table on which to record the answers to the questionnaire.

Questionnaire

1 Are you in this area for the purpose of shopping? (If 'No', then discontinue the interview.)
2 Which area of this town do you live in?
3 Which of these items do you usually buy at this centre?
  (a) food
  (b) medicines and other items from the chemist
  (c) clothes
4 How did you reach this shopping centre?
  (a) by foot
  (b) by bus
  (c) by car
  (d) other
5 How often do you visit this shopping centre?
  (a) every day
  (b) two to five times a week
  (c) once a week
  (d) less than once a week

N.B. The three items chosen for question 3 should represent low order, middle order and high order goods.

Processing data

Analyse the answers to the questionnaire carefully, making sure that you keep the results from each shopping centre separate.

Answers to questions 2 and 3 must be kept together and tabulated (see table 7.6). This information will be needed

7.6 Questionnaire for (named) shopping centre

Interviewee	Area of town lived in (Qu 2)	Food (Qu 3a)	Items from the chemist (Qu 3b)	Clothes (Qu 3c)
1	Exeville	✓	✓	✓
2	Wyeville	✓		
3	Zedville	✓	✓	
etc.				

for mapping the trade areas.

Answers to questions 4 and 5 can simply be recorded as totals. The raw data should then be converted to percentages as this will enable comparison between the centres even if different numbers of interviews have been completed (see table 7.7).

Presentation of results

1 Map the shopping centres chosen to show the number, layout and types of shop in each centre. Devise a colour key and use the same key for each shopping centre. Alternatively, three colours could be chosen to distinguish shops selling low, medium and high order goods.

2 Count the number of shops in each centre in each category (food shops, clothes and shoe shops, commercial services etc.). Present the information using bar charts. Make sure that the same scale and key is used for each centre so that they can be compared.

3 Map the trade area of each shopping centre. You will need to prepare a base map which shows all the shopping centres being studied and the areas of the town where most people live. A town plan or 1 : 50,000 OS map is useful for this. Use a separate map for each shopping centre and plot desire lines to show the area from which customers come (see fig. 7.8). Different colours could be used to show whether the customers use the centre for food, the chemist, or clothes, using the information obtained in Question 3 of the questionnaire.

4 Draw graphs to show the methods of transport used to get to each centre. Remember to use the same method and key for each centre.

5 Draw graphs to show how often customers visit each shopping centre. Again the graphs must be comparable for all the centres but try not to use the same type of graph you used for the methods of transport.

Divided column graphs or proportional circles could be used to show either transport or frequency of visits (see Chapter 1).

7.7 Methods of travel used by shoppers at 3 different centres

	Shopping centre 1		Shopping centre 2		Shopping centre 3	
	Number	%	Number	%	Number	%
Foot	25	50	15	25	18	33
Bus	12	24	30	50	11	20
Car	13	26	13	22	25	45
Other	0	0	2	3	1	2
Total	50	100	60	100	55	100

7.8 Desire line links showing the trade areas of three shopping centres in west Sheffield

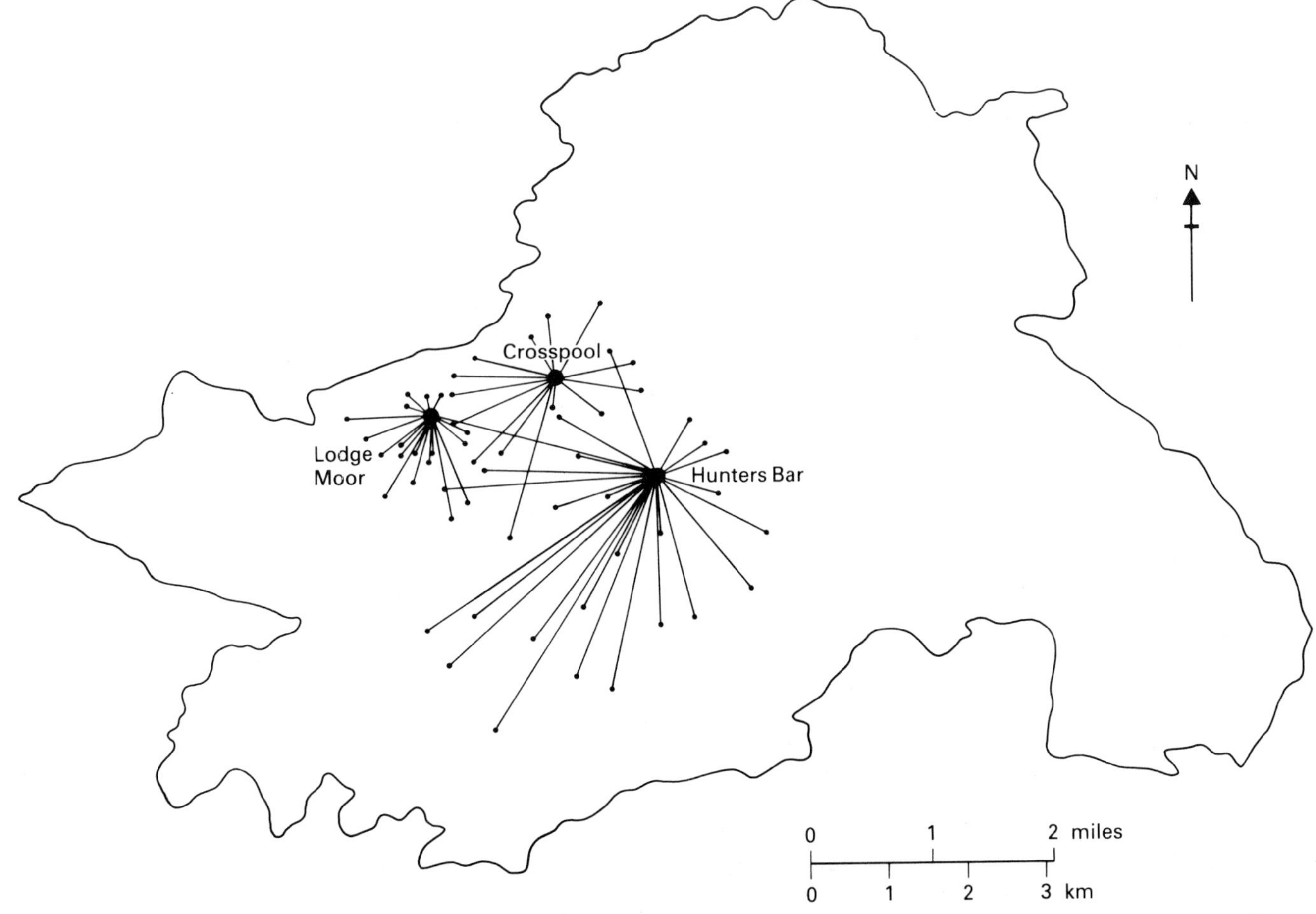

Analysis

Your results should provide plenty of information about the characteristics of the shopping centres themselves and about the ways in which they are used by customers. It is important to discuss your results in a clear and logical order. First try to describe and explain the differences in the size of the centres. You should think particularly about the orders of goods being sold in each centre.

Secondly, compare the trade areas of each of the shopping centres. You may find a considerable difference in both shape and size and you should attempt to explain this. Remember the position of the centre in relation to main roads and bus routes will affect the size and shape of the trade area as will the density of housing in the area, the location of competing shopping centres and of course the shops and services available in the centre itself.

Thirdly, you should compare the methods of transport used by customers visiting each shopping centre, and the frequency of visits. Again you may find considerable differences between the shopping centres which you must attempt to explain. On the other hand there may not be great differences between the centres and you must be sure to describe and explain the patterns you find and not the patterns you hope or expect to find.

Finally, you must draw your work together and make some conclusions. Remember to refer back to your initial hypothesis which you may wish to accept or reject. You may also be able to suggest ways in which your fieldwork could be improved.

Suggestions for further study

This type of fieldwork can be used to study the character and trade area of many things, such as:

1 The character and trade area of a market town.
2 Differences in the trade area and customers' habits between a superstore on the outskirts of a town and an established suburban shopping centre. Alternatively you could look at the impact of a new superstore, and its effects on nearby shopping centres.
3 Comparison between trade areas of contrasting leisure facilities, e.g. a children's playground, a bingo hall, a skating rink and a theatre.
4 An interesting extension to this study is to use W.J. Reilly's Gravity Model to establish the theoretical breaking point between shopping centres and to test this against the actual breaking point obtained by questionnaire.

$$\text{Breaking point from centre B} = \frac{\text{Distance between centre A \& centre B}}{1+\sqrt{\left(\dfrac{\text{Size of centre A}}{\text{Size of centre B}}\right)}}$$

The size of the shopping centres can be established by counting shops (or by looking at population size if different towns are being studied). You should calculate the breaking point between the shopping centre and all the surrounding centres, and use this to establish the trade area of the centre, as shown in fig. 7.9.

Further reading

K Briggs, *Fieldwork in Urban Geography*, Oliver & Boyd 1971

J A Everson and B P Fitzgerald, *Settlement Patterns*, Longman 1969

P Toyne and P Newby, *Techniques in Human Geography*, Macmillan 1971

7.9 Theoretical trade area of a shopping centre

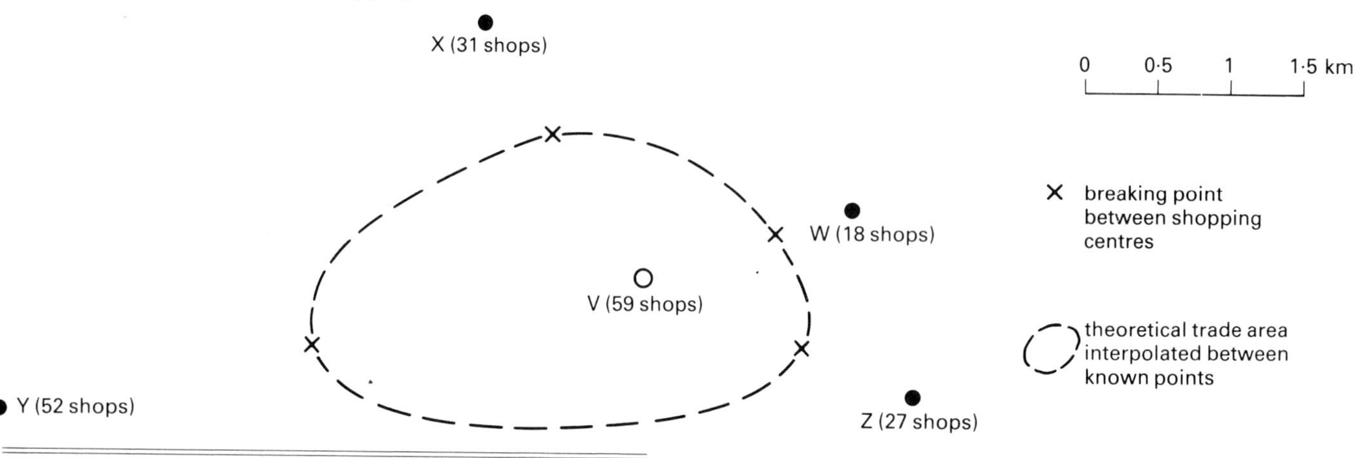

Centre	No of shops	Distance from V	Breaking point from V
W	18	1.5 km	1.5 ÷ 1.55 = 0.97 km
X	31	2 km	2 ÷ 1.72 = 1.16 km
Y	52	4.5 km	4.5 ÷ 1.94 = 2.32 km
Z	27	2 km	2 ÷ 1.68 = 1.19 km

8 Variations in soil profiles

Suitable for individual or group study

Topic for study

Soil is the product of interaction between the inorganic (geological) and organic (biological) parts of the natural system. It also contains air and water in amounts which vary inversely. Any changes in input from the inorganic or organic parts of the natural system will be reflected in the composition and characteristics of the soil.

The inorganic fraction consists of particles of sand, silt or clay which have been released from the parent material by weathering. The organic fraction consists of humus (decaying remains of the plant cover) which may be incorporated into the soil horizons or which may form surface layers. If the weathered material contains clay particles these will combine with the humus to produce the *clay-humus complexes* (or micelles), which exert a negative charge. The micelles attract freely moving positively charged calcium, magnesium, potassium, ammonium and sodium ions and thereby are responsible for the soil's *fertility*. However, if the soil is frequently subject to wetting by downward percolating rainwater, hydrogen ions (which are released as the water dissociates into hydrogen and hydroxide ions) replace the calcium etc. around the micelle in a process called *leaching*. The extent to which leaching has occurred can be measured on a pH meter or kit; pH values range from 1 to 14. Very acid soils have pH readings of 4 and alkaline ones of 8 or 9. A further stage in the leaching of the soil is called *podsolisation* which is the process whereby the clay particles themselves break up into the three constituents of iron, aluminium and silica. A soil which has undergone this process has distinctive horizons of different colours.

Consequently a study of soil should involve the measurement of the following variables:

1 pH
2 moisture content
3 particle size to determine the proportion of fines (small particles) to coarse fragments
4 soil depth and position on slope

Profiles should also be drawn. The number, arrangement and colour of horizons indicate the soil processes operating at that site. (For more detail on soils, consult Bridges, *World Soils*.)

Possible hypotheses

1 Changes in moisture content, particle size and humus characteristics produce distinctive soil profiles.
2 As the angle of slope increases the depth of the soil decreases as does the moisture content.
3 As moisture content increases so does soil acidity.
4 Changes in vegetation affect soil acidity.
5 Particle size affects moisture retention.
6 Slope position affects profile depth.

Location

It is interesting to investigate changes in soil profile downslope. Changes in slope angle and slope position might be expected to affect moisture content and soil depth. Choose a valley side on which geology, climate and aspect are constant but where there are obvious changes in slope angle and vegetation. You *must* obtain permission to dig. You will need 10 sites if you are to use Spearman's test in your analysis of results and 20 sites if you are to use the chi-square test. You may decide to collect data from several transects down the same slope.

Data collection

Having selected a suitable slope (with reference to a geology map so that you keep this variable constant) decide on an appropriate method of sampling (see Appendix A). You should endeavour to investigate the soil wherever there is a marked change in vegetation.

Equipment
30-metre tape
clinometer
metre rule
spade
large sheet polythene (e.g. dustbin liner)
small labelled plastic bags for soil samples (at least 3 per site)

In the laboratory
pH kit/meter
set of soil sieves
small paintbrush
pestle and mortar
calipers
scales
pocket calculator

Method

1 Stretch tape at right angles to the slope and conduct a slope survey using the clinometer (see chapter 6). You will need to devise a slope recording sheet.
2 Use your sampling method to select each site location. Record the position of each site on the slope recording sheet.
3 At each site
   (*a*) Slice through and under the vegetation and humus on three sides of a 1 metre square, then roll back the carpet of vegetation
   (*b*) Dig a pit 1 metre square, keeping the sides as clean as possible until you encounter *either* the bedrock *or* the water table. Place all the soil you have removed on the large polythene sheet so that you can replace it afterwards.
   (*c*) Using the metre rule measure the depth of each horizon and record on a prepared sheet (fig. 8.1).
   (*d*) Take a small soil sample from each identifiable horizon. Label with site *and* horizon. For information about horizons, their arrangement, texture, structure and colour see Bridges, *World Soils*. You will certainly find an A horizon and probably several others beneath it.

8.1 Soil profile sheet

Site ________________

Slope position (top, middle, bottom) ________________

Slope angle ________________

Vegetation type ________________

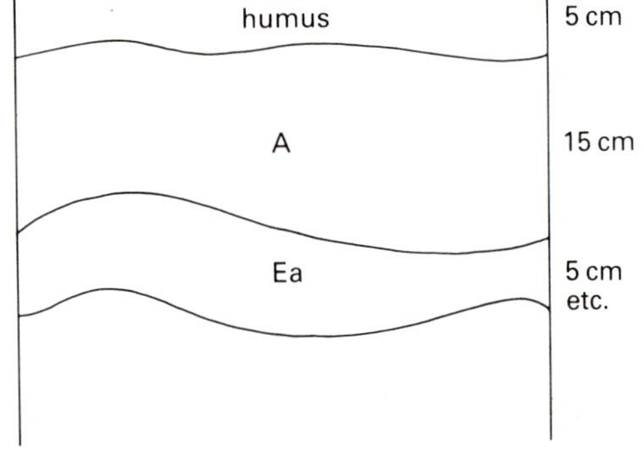

The sample you take from the A horizon should be larger than from the other horizons as you are going to test for particle size.

Processing data

1 Test a small amount of each sample for pH. Record on a combined results sheet (fig. 8.2).
2 The size of particles present in the soil is partly the result of weathering characteristics of the bedrock and its composition and partly the result of downslope movement of particles. Test the A horizon for particle size as set out below. The percentage of fines (very small particles) will be significant as these include clay particles of less than 0.02 mm in diameter which combine with the humus to form clay–humus micelles.

8.3 Particle dimensions

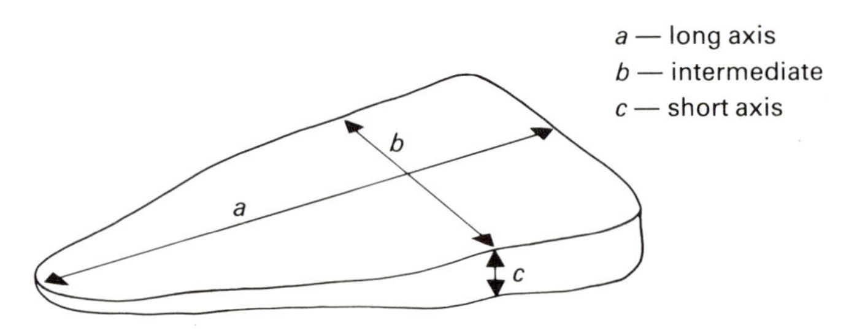

8.4 Separating out the soil samples

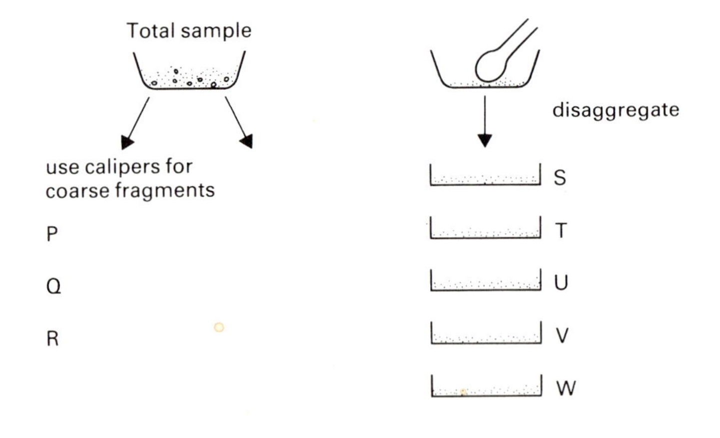

8.2 Combined results sheet

	Environmental characteristics			Soil characteristics																				
				pH										Moisture %										
				Individual horizons									mean for site	Individual horizons									mean for site	% finest particles
Site	Slope angle (°)	Slope position	Vegetation	H	A	Ea	Eag	Bh	Bfe	Bs	Bg	C		H	A	Ea	Eag	Bh	Bfe	Bs	Bg	C		
1	2	top	ling	5	5	4.5	–	4.5	4.5	5	–	5	4.8	25	22	19	–	21	15	19	–	19	20	29
2																								
3																								
etc.																								

Method of determining particle size

(a) Weigh soil sample in a small tin or evaporating dish. Record this on a soil results sheet (fig. 8.5).

(b) Set the calipers at 10 mm then pick out all fragments with an intermediate axis greater than 10 mm.

(c) Brush off the fine material sticking to the coarse fragments. Do not lose it – put it with the fine particles.

(d) Split the coarse fragments into the following sizes using the calipers:

over 40 mm P
20–40 mm Q
10–20 mm R

Weigh each category and record the weights on the sheet. (Remember to weigh each *tin* first.)

(e) Use a pestle and mortar to disaggregate the particles which are finer than 10 mm. Do *not* break up individual

8.5 Soil results sheet

Site________________ Total weight of sample ____________

	Caliper or sieve gauge (mm)	Tin no	Weight of tin	Weight of tin & sample	Weight of sample	Sample weight as % of total weight	Material finer than sample as % of total weight
P	> 40 mm					5	95
Q	20–40 mm					10	85
R	10–20 mm					2	83
S	etc.					7	76
T						20	56
U						15	41
V						12	29
W						29	0

8.6 Diagrammatic slope profile

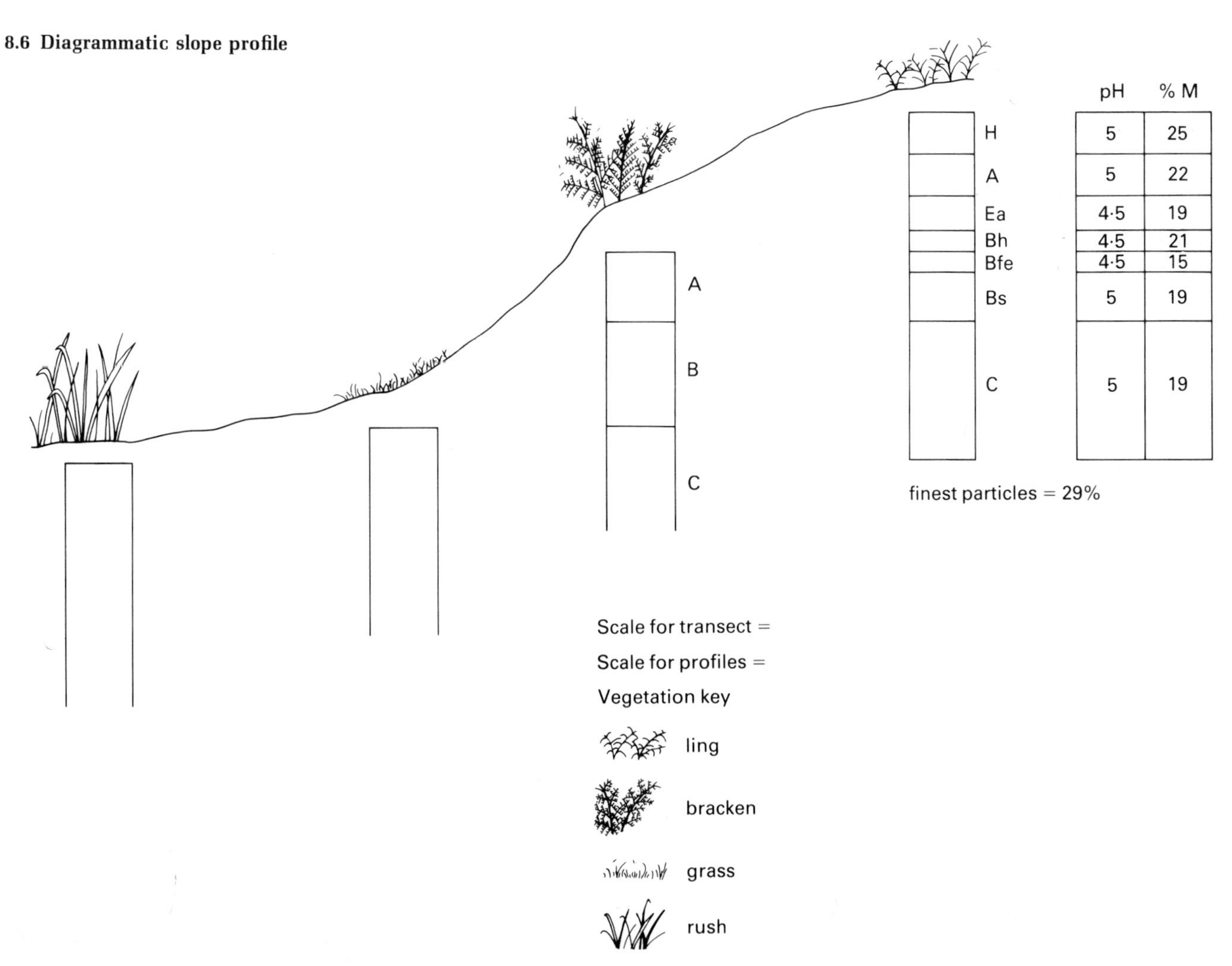

particles but make sure soil clods are crushed.

(*f*) Put the fine materials through the set of sieves. Shake the set of sieves vigorously. Weigh the contents of each sieve S–W, remembering to weigh the tin into which you put the contents first. Record the actual weights on the results sheet (fig. 8.5). Each sieve will be labelled with the gauge size. Record this in the size column.

(*g*) If we are to compare the sizes we need to convert each weight into a percentage of the total weight. Use the formula:

$$\text{percentage weight} = \frac{\text{wt of fraction}}{\text{wt of total sample}} \times 100$$

Record this information on the soil results sheet (fig. 8.5) and enter the percentage weight of the finest material (W) in the combined results sheet (fig. 8.2).

We also need to know what percentage remains after the coarser fragments have been removed. This is calculated as follows: if 5% of the sample is *greater* than size P, 95% must be *smaller* than size P. Record these values on the soil results sheet.

3 Test for moisture content for each horizon (see chapter 4 for method).

Presentation of results

1 On a large sheet of paper draw out the slope profiles to scale, leaving enough space between two profiles to insert a scale drawing of the soil profile. Insert information about vegetation, pH, moisture content and fines in an appropriate manner. One method of representation is shown in fig. 8.6. Instead of labelling each horizon you may prefer to use a colour code and make a key.

2 Test hypotheses 2 and 3 using either scatter graphs or Spearman's rank test. If you choose scatter graphs remember to put the independent variable on the x axis.

3 Test hypotheses 4 and 6 using dispersal diagrams or the chi-square test.

Dispersal diagrams are ideally suited as you are comparing a quantified variable and a non-quantified variable in each hypothesis. Use the actual pH values for a dispersal diagram (fig. 8.7).

8.7 Dispersal diagram relating vegetation to soil acidity

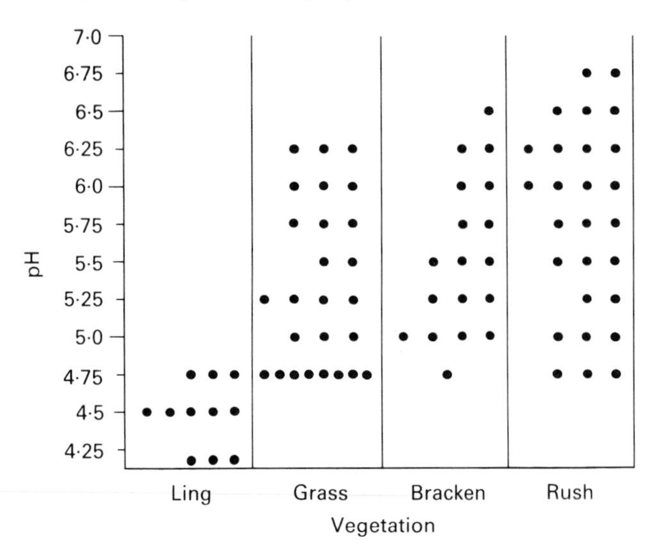

If you intend to use the chi-square test you must check your data against the requirements of the test before using it.

(*a*) There must be at least 20 observations.

(*b*) The data must be in the form of frequencies. Percentages must *not* be used.

(*c*) No expected frequency must be less than 1 and not more than 20% of the expected frequencies must be less than 5.

(*d*) The observations must be independent.

For the method of using the test see chapter 12. Set out your results in a *matrix* as shown in fig. 8.8 where pairs of entries are observed data and, in brackets, expected data.

8.8 Vegetation–acidity matrix (observed and (bracketed) expected data)

Veg. Type	Soil acidity			Total
	4–4.9	5–5.9	6–6.9	
Ling	11 (5.1)	11 (11)	0 (5.9)	22
Grass	8 (6.1)	12 (13)	6 (6.9)	26
Bracken	1 (4.2)	12 (9)	5 (4.8)	18
Rush	1 (5.6)	10 (12)	13 (6.4)	24
Total	21	45	24	90

Null hypothesis (H_0). There is no significant difference in the pH values for ling, grass, bracken and rush.

χ^2 for the above data is 28.81 and df=6. Therefore by looking at the significance tables (see Appendix C) we can reject our null hypothesis at the 0.01 confidence level and put forward the hypothesis: There is a significant difference between the pH values of ling, grass, bracken and rush.

4 A simple pyramid graph may be used to test hypothesis 5. Use the mean percentage moisture content for each site.

Site	Finest particles (%)	Moisture (mean %)
1	29	20
2	15	25
3	50	35
4	10	5
5	26	17

8.9 Pyramid graph relating moisture to fine particles

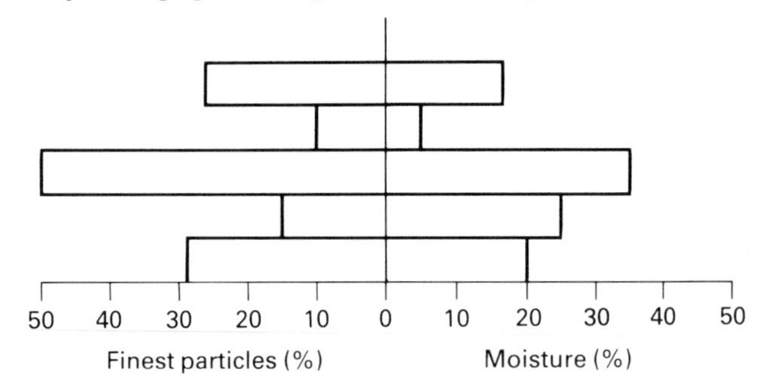

5 Using the data from each soil results sheet (8.5) plot a cumulative frequency graph for each site. This type of graph enables us to see clearly the texture of the soil. Comparison of the cumulative frequency graphs, which could be superimposed for each transect enables us to see changes in texture down slope.

8.10 Cumulative frequency graph showing variation of particle size

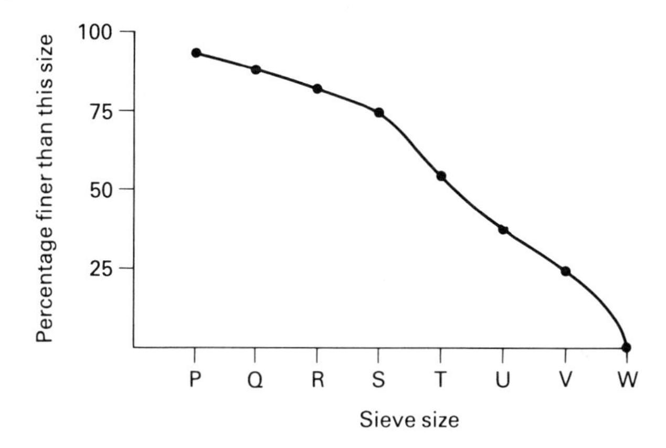

Analysis

Your diagrammatic slopes should have revealed all the major changes in soil and environmental characteristics and you should therefore be able to *describe* the changes accurately. However, hypothesis 1 anticipates a discussion of *why* these changes have been responsible for producing different soil profiles. You may therefore find it more relevant to discuss your results from hypotheses 2–6 before you attempt to explain the downslope variations.

If you have made several transects it may be worthwhile examining the range of conditions present at different slope positions or beneath different vegetation types.

Suggestions for further study

1 Attempt to relate particle size to slope position and investigate possible mass movement processes operating on the slope.
2 A comparison of soil types on differing parent materials could be made.

Further reading

E M Bridges, *World Soils*, 2nd edn, CUP 1978
P McCullagh, *Data Use and Interpretation*, OUP 1974

9 Infiltration in an upland region

Suitable for group study

Topic for study

The infiltration rate is the rate at which soil absorbs water, measured in mm/minute. Eventually infiltration reaches a constant rate, known as the *infiltration capacity*. This represents the soil's ability to absorb water.

Infiltration is a major process operating within a drainage basin. It is part of the system in which rainfall (or precipitation) reaches the river channels and becomes run off.

When precipitation encounters the ground it may

1 be intercepted by leaves of plants,
2 lie on the surface temporarily prior to infiltration,
3 infiltrate immediately,
4 run straight off into rivers and streams if the ground is already saturated,
5 lie in depressions or hollows if the ground is already saturated.

The rate at which the moisture infiltrates depends on:

1 *Preceding weather conditions.* If the weather has been hot and dry the ground will be baked hard and the rainfall will initially collect on the surface before infiltration. Persistent rain on the other hand will have made the ground very wet and pores will be open. Infiltration will continue until saturation is reached, when it ceases and overland flow will commence.
2 *Slope position.* A downslope site is more likely to become saturated than an upslope site as it receives infiltration of its own *plus* that which is part of the throughflow, from upslope. Infiltration rates and capacity will thus be low and saturation occurs more quickly.
3 *Soil texture and structure.* Sandy soils have large pore spaces and a crumb structure hence infiltration is usually rapid. Clayey soils have tiny pores and a platey structure so moisture often collects on the surface before infiltration commences.
4 *Slope angles.* Steep slopes encourage run off rather than infiltration, and the soil will have a low moisture content. Hence, although they have dry soils, their *ability* to absorb water is high should infiltration commence.

Hypotheses

1 Lower infiltration rates and infiltration capacity occur near the bottom of the slope.
2 Steep slopes have a high infiltration capacity (ability to absorb water).
3 The higher the percentage of sand the higher the infiltration rate.
4 A change in weather conditions influences infiltration rates at the same site. (If you intend to test this hypothesis you will need to repeat the whole experiment after a change in weather conditions.)

Location

Choose a valley side with a variety of slope angles within 100 metres of a stream. You will need access to the stream for water during the experiment. You must keep vegetation type fairly constant as different root systems and density of plant cover affect infiltration rates. Remember to obtain permission if you intend to carry out the experiment on private land. (For layout of sites see fig. 9.4.)

9.1 Infiltration in an upland region

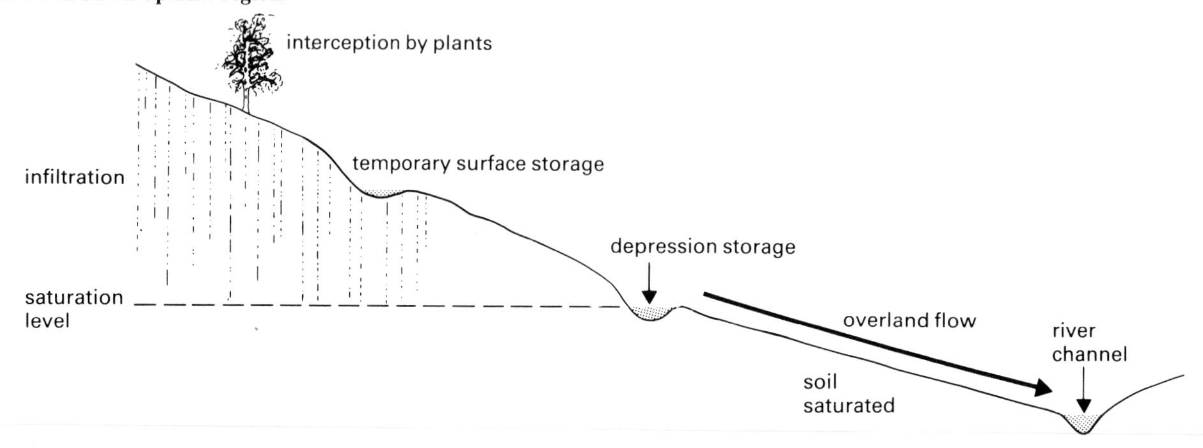

9.2 The effect of slope position on infiltration

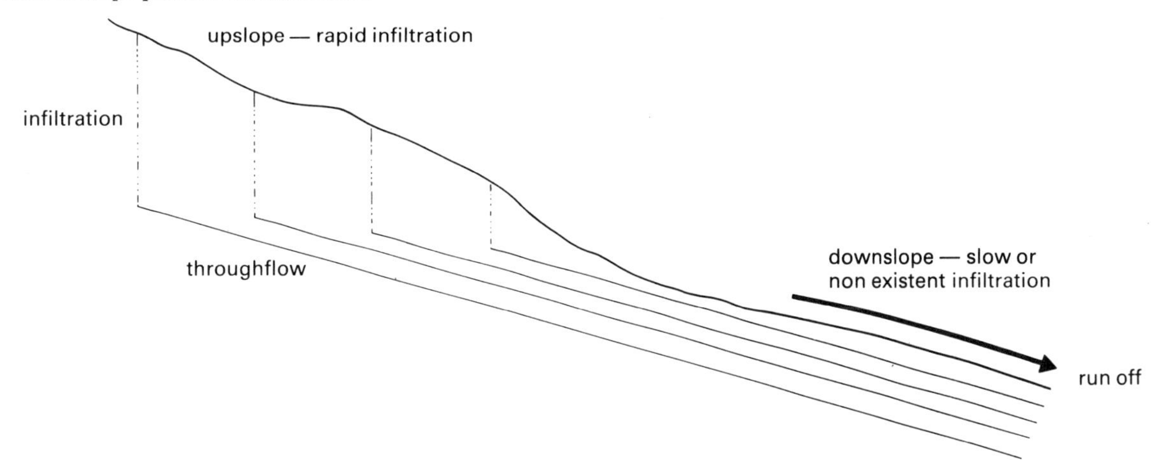

9.3 Home-made clinometer

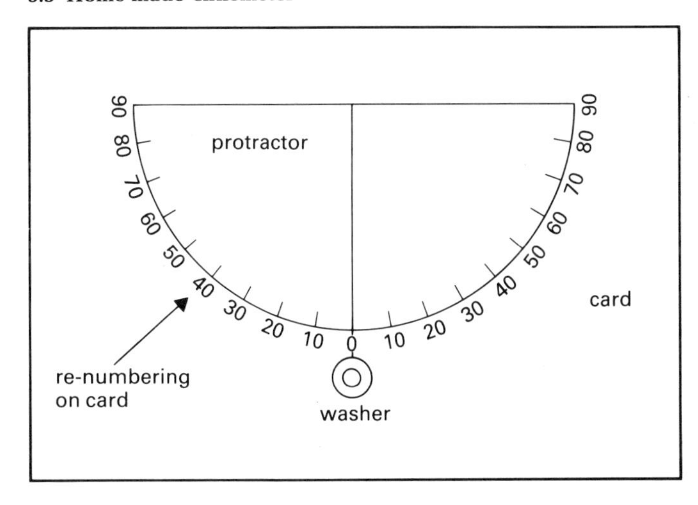

9.4 Sampling sites

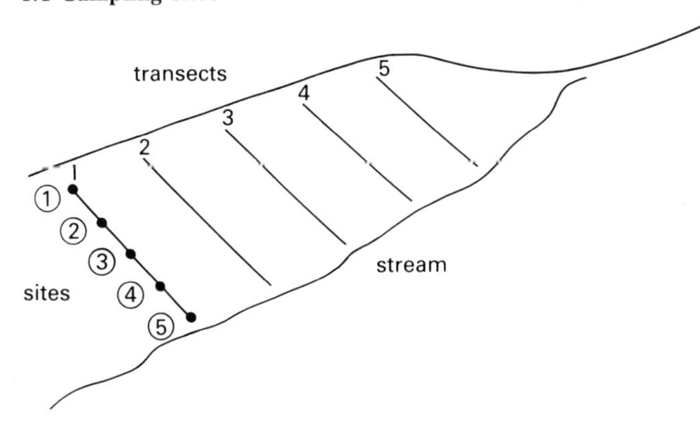

9.5 Clinometer reading

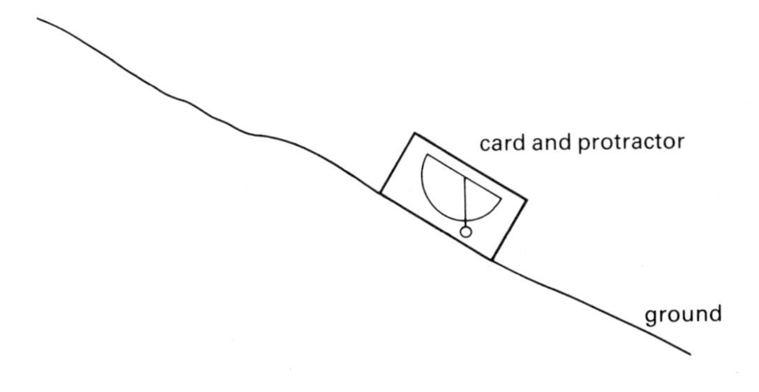

Data collection

Advance preparation

You will need to record information on five transects, so prepare five recording sheets (fig. 9.6).

Equipment

Small cans (large dog food/catering tin) with top and bottom removed to make a tube; the cans soon become unusable so take at least 10 with you
plastic ruler with scale in millimetres
30-metre tape
stop watch/digital watch with minutes
trowel and rubber headed hammer (such as used for tent pegs)
plastic bucket(s) for collecting water
ready-labelled plastic bags for each site
clinometer which can rest on the ground; you may need to devise your own with a piece of card, protractor, string and metal washer (fig. 9.3)
set of soil sieves/pestle and mortar

Method

1 Stretch the tape upslope at right angles to the stream, beginning $\frac{1}{2}$ metre in from the bank. You should aim to select five sites per transect at regular intervals. Select a distance interval to suit all your transects (fig. 9.4).
2 Using the trowel, cut round a circle corresponding to the can's circumference. Insert the can to a depth of 5 cm, knocking it in with the hammer. Try to disturb the vegetation and soil as little as possible. Keep the can upright even when the slope is steep.
3 Put the ruler into the can making sure that it will remain upright when the can is filled with water.
4 Place the clinometer on the ground next to the can and take a slope reading as shown on fig. 9.5.
5 Fill the bucket, then fill the can, replenishing as necessary.
6 Time the fall of the water in mm per minute for *at least* 15 minutes. A more accurate result will be obtained if you read over 30 minutes, but this is very time-consuming unless this exercise is undertaken by a group.

7 When you have finished take a soil sample (for testing the percentage of sand) from beneath the can.
8 Repeat for four more sites on the first transect and then complete four more transects on the same side of the stream.

9.6 Infiltration recording sheet

Time	Transect no ______				
	Site 1	Site 2	Site 3	Site 4	Site 5
	Angle of slope (°)				
	8	12	7	3	1
	% sand				
	80	80	72	63	60
(min)	Fall of water (mm)				
1	20	15	10	4	2
2	18	10	9	4	2
3	13	9	6	4	1
4	9	5	4	3	1
5	8	5	4	3	0
6	8	5	4	3	0
7	7	5	4	2	0
8	7	5	4	2	0
9	6	5	4	2	0
10	6	4	4	1	0
11	4	4	3	1	0
12	4	4	3	1	0
13	4	4	3	0	0
14	4	4	3	0	0
15	4	4	3	0	0
	Total fall (in 15 min)				
	122	88	68	30	6
	Infiltration rate (mm/min)				
	8.1	5.9	4.5	2	0.4
	Infiltration capacity (mm/min)				
	4	4	3	0.4	0

9.8 Rate of fall (data for transect 1)

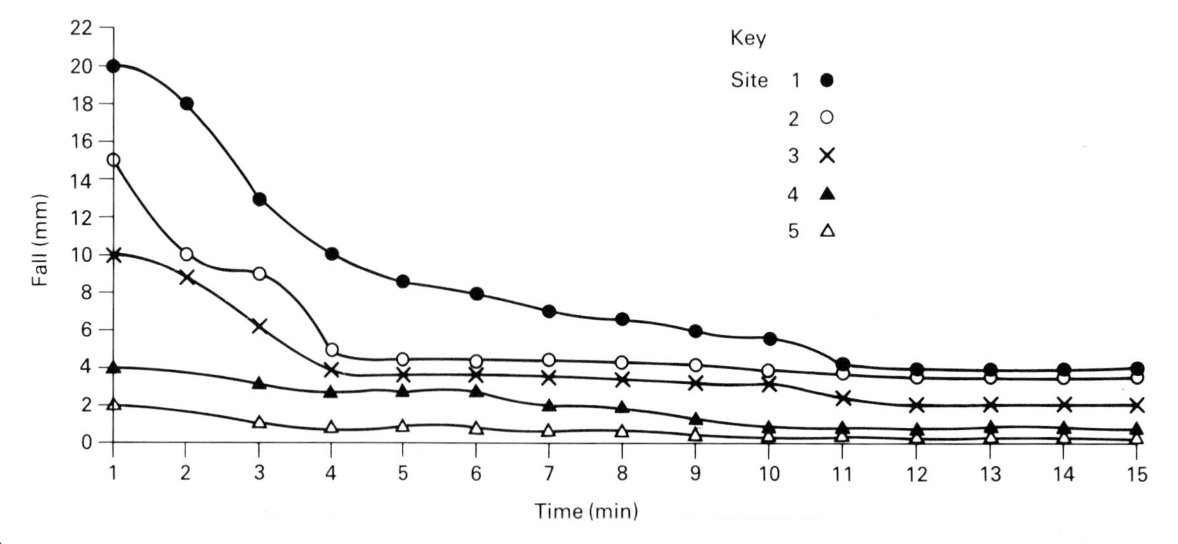

Processing data

When all the readings have been taken, calculate:

(*a*) Total fall over 15 minutes
(*b*) Mean fall (mm/minute) – the *infiltration rate*
(*c*) Infiltration capacity – when the fall has become constant the *infiltration capacity* has been reached. The last five readings can be used to calculate infiltration capacity. The infiltration capacity can be converted to cm/hour.
(*d*) Put the soil sample through the sieves and test for percentage sand (particles over 0.2 mm in size). See chapter 8 for method.
(*e*) Prepare a combined results sheet as shown in table 9.7 which includes processed data from table 9.6.

Presentation of results

1 Prepare a graph for each transect. Using a different colour or symbol for each *site*, plot fall in mm against time (fig. 9.8).
2 Draw diagrammatic slopes for each transect beneath one another on one sheet of paper. Draw the shape of the slope and mark the position of each site. The slope will not be drawn to scale but you should space out the sites evenly to show that you made a systematic survey.

9.7 Combined results sheet

Transect		Site 1	Site 2	Site 3	Site 4	Site 5
1	I rate (mm/min)	8.1	5.9	4.5	2	0.4
	I capacity (mm/min)	4	4	3	0.4	0
	Angle (°)	8	12	7	3	1
	% sand	80	80	72	63	60
2	I rate					
	I capacity					
	Angle					
	% sand					
etc.						

9.9 Diagrammatic slopes (data for transect 1)

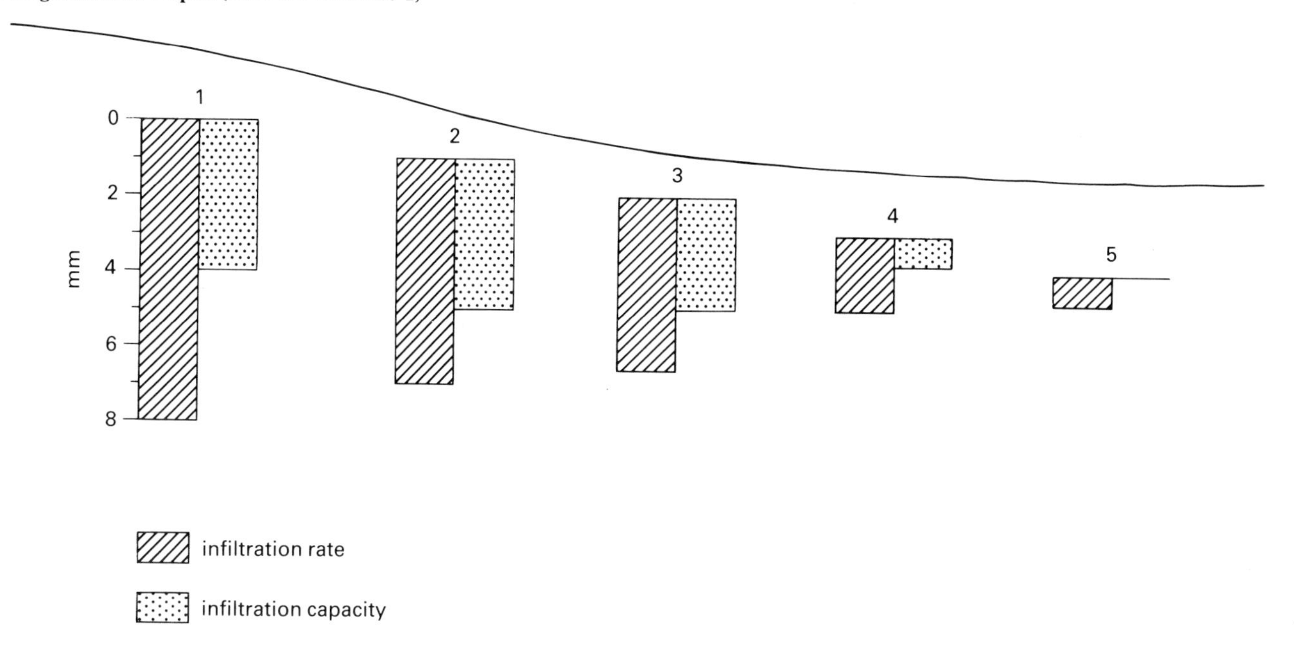

Beneath each site construct two columns as shown in fig. 9.9 to represent the infiltration rate and infiltration capacity. Choose a suitable scale for your data.

3 Draw a scatter graph to test the hypothesis: 'The steeper the slope the higher the infiltration capacity'.
Plot each transect in a different colour. Put slope angle (in degrees – the independent variable) on the x axis and infiltration capacity on the y axis.

4 *Either* draw a scatter graph *or* use Spearman's rank correlation to test the hypothesis: 'The higher the percentage of sand the higher the infiltration rate'. If you draw a scatter graph put the independent variable (sand) on the x axis and infiltration rate on the y axis.

5 If you have carried out the experiment twice in order to test hypothesis 4 set out your results in a *contingency table* (9.11).
Either draw two histograms to represent the data or use the chi-square test. Remember to set a null hypothesis: 'There is no difference between infiltration rates after wet or dry weather'. Check that your data are suitable for this test (see chapter 12).

9.10 Scatter graph showing the relationship between slope angle and infiltration capacity

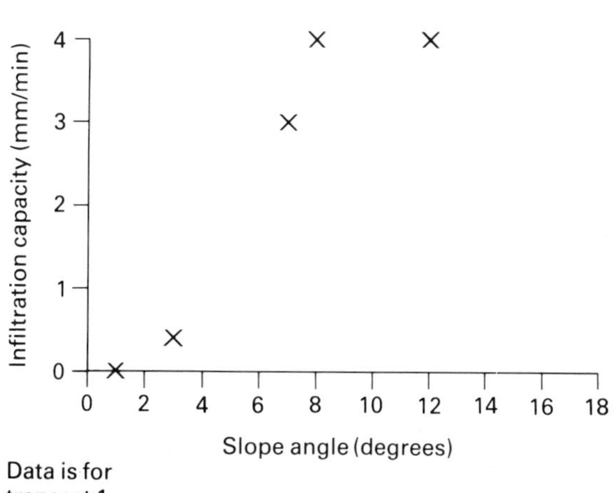

9.11 Contingency table showing infiltration rates after different weather conditions

Weather type	Infiltration rate					
	0–2	2.1–4	4.1–6	6.1–8	8.1–10	Total
Wet weather	7	6	7	1	4	25
Dry weather	1	4	5	7	8	25
Total	8	10	12	8	12	50

Analysis

The aim of the discussion should be to find reasons for the results you have obtained. Your diagrammatic slope should enable you to accept or reject hypothesis 1. You may find very high initial infiltration rates followed by an immediate slow down producing a high mean rate (see transect 1, sites 1 and 2). This initial high rate may be due to the creation of a 'head' of water as the water is first poured into the tin. Very rapid rates throughout your 15/30 minutes could mean that sideways seepage is taking place at the base of the tin. Be aware that faulty experimental technique may lead to a need to reject the hypothesis. When saturation is reached the infiltration capacity will be 0 e.g. transect 1 site 5. Discuss the extent to which saturation has occurred in your results.

Hypothesis 2, if proved, will show a positive correlation between steep slope and infiltration capacity. Should you find no correlation between slope angle and infiltration capacity, consider the extent to which slope position (as

9.12 Histograms showing effect of previous weather on infiltration rates

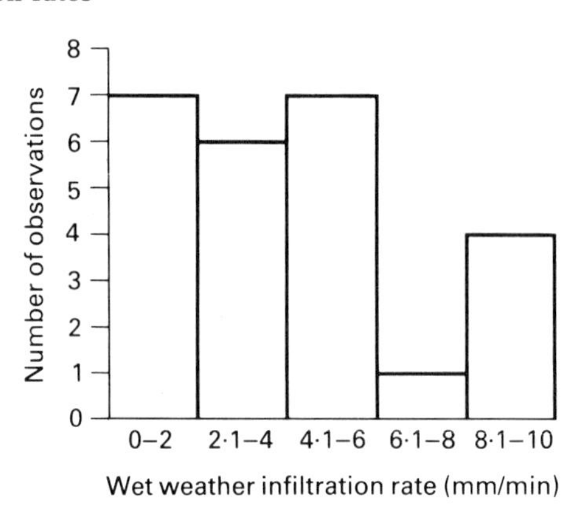

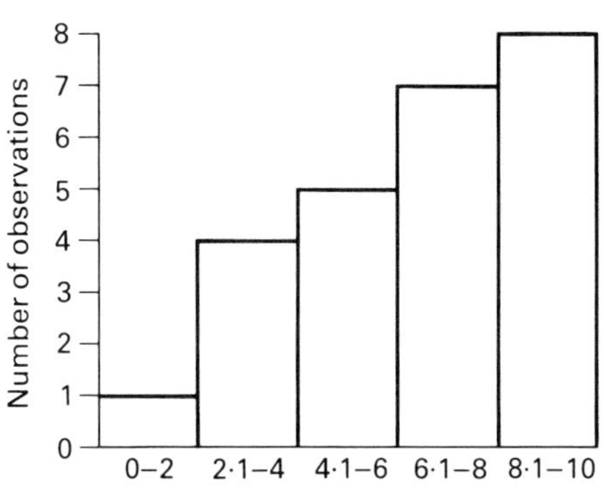

shown on the diagrammatic slope) is more important than slope angle.

Hypothesis 3 may not prove positive for the same reason – slope position may be more important than soil texture/structure. There may be little variation in soil texture throughout the area, though normally the fines (see chapter 8) will have washed downslope by mass movement/overland flow.

When considering Hypothesis 4 remember that *very dry* antecedent weather conditions can lead to low infiltration rates as the ground will be hard and compact. Try to give an explanation for the results you have obtained.

Suggestions for further study

1 Prior to inserting the tin take an adjacent soil sample and test for moisture content (see chapter 4). See whether or not there is a correlation between percentage moisture and infiltration rates/capacity.
2 Find an area with different vegetation types on the same slope position. Find out how infiltration rates vary under different vegetation types.
3 Conduct the experiment in two geologically different areas.

Further reading

E M Bridges, *World Soils*, 2nd edn, CUP 1978

J D Hanwell and M D Newson, *Techniques in Physical Geography*, Macmillan 1973

10 Transport studies

Suitable for group study

Topic for study

Studying traffic flows or bus and rail services is a fairly simple procedure yet it can yield useful information about settlements and may help to explain why some have grown at the expense of others. The results may reveal differences in access by public transport compared to private transport and this may be reflected in patterns of journey to work. On the other hand the results may serve to emphasise the influence of relief, or conversely that physical factors are not of major importance. Thus a thorough study can provide a great deal of geographical information about a particular area.

Possible hypotheses

A number of different short hypotheses is a good approach to this area of study.

1 The interaction statistic can be used to predict traffic flow between settlements.
2 The bus service between settlements is related to the size of settlement served and the length of journey.
3 Natural features influence the time taken to travel between settlements.

You could devise a fourth hypothesis based on cost of travelling between settlements.

Location

Choose a small or medium-sized town that is linked to several other larger and smaller settlements by road. The central town must have a bus station and provide bus services to the settlements in the surrounding area. Visits to the bus station itself will be needed in order to collect information about numbers, times and cost of bus journeys. Large towns or cities are probably best avoided because of the huge numbers of different traffic routes and bus services.

Data collection

In this study you are testing a number of different hypotheses and therefore you must be certain to collect all the data required for each hypothesis. Before collecting any data decide which settlements in the surrounding area you are going to include in the study. Choose between four and eight settlements up to a maximum of 25 miles away. The centre you have chosen as a base for your study should be roughly in the middle of these selected settlements.

Look at OS maps, and road maps, and study population figures to help you to decide. The *AA Members' Handbook* is a useful source of information.

Secondly, you must consider the number of people available to collect data. If you need to collect data to compare traffic flows on different routes (hypothesis 1) then there must be enough people to count traffic on each of the routes at the same time.

To test the hypotheses suggested above proceed as follows:

1 *Traffic flow*: Record traffic flow between the central settlement and each of the selected surrounding settlements. Sometimes two settlements are reached by the same route and in this case it will be impossible to differentiate between traffic going to these settlements. This does not matter.

   The count must be taken from positions beyond road junctions (see fig. 10.1) and must be taken on all the routes simultaneously. Count and record vehicles passing in both directions over a period of at least one hour, and preferably longer.

2 *The interaction statistic*: Collect the following information which is required to calculate this statistic:
   (*a*) population size of the selected settlements,
   (*b*) distance in miles (or km) between the central settlement and each selected surrounding settlement

10.1 Traffic flow to the surrounding settlements should be counted at points A to F

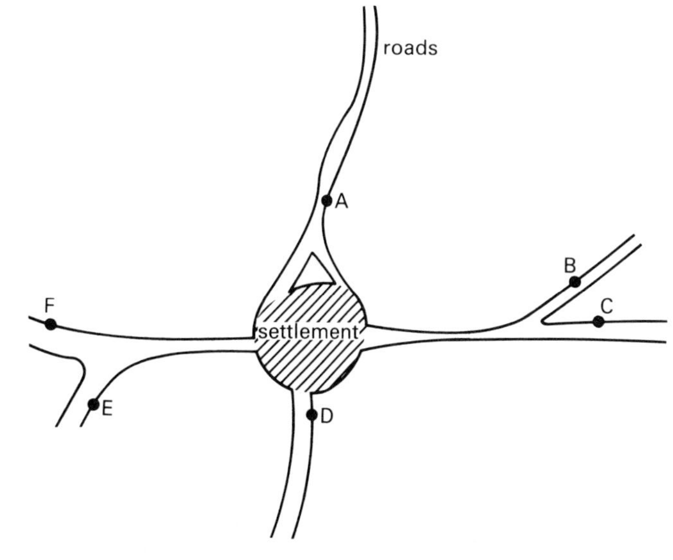

The *AA Members' Handbook* is a useful source of information.

3 *Bus services*: Find the number of buses running between the central settlement and each of the selected settlements. The best way is to visit the local bus station and consult the timetables. Alternatively buy a copy of the published timetables and take it away for a more leisurely study. Remember that the number of buses may vary during the year and indeed on different days of the week, therefore it is important to decide which information you are going to collect. Make certain all the information you collect is comparable, for example, take the number of buses per week in summer.

4 *Times of bus journeys*: Find as many settlements as possible within your chosen area which can be reached by bus. Record the time taken in minutes to reach each of these places by bus from the central settlement. *At least* 30 journeys must be recorded. You may find that some buses take longer than others and therefore you must decide whether to record the average time, or the fastest time or the modal time (most commonly occurring time). Be careful, do not write down the first time you discover.

This is another area of data collection that can be done more carefully if you buy a book of bus timetables, which you can take home and study.

5 *Cost of bus journeys*: Find out the cost of travelling to all the settlements in the area. Bus companies usually publish lists of fares and this should provide the information required. At least 30 journey fares should be recorded. This will only be necessary if you have devised a hypothesis concerning the cost of journey.

10.2 Interaction statistic between Ambleside (population 2,657) and surrounding settlements

Settlement	Population in thousands*	Miles from Ambleside*	Interaction statistic
Grasmere	.990	4	16.44
Hawkshead	.684	5	7.27
Windermere	7.860	5	83.54
Kendal	22.440	13	35.28
Keswick	4.790	17	4.40
Coniston	1.063	8	4.41

*Population and mileage figures from *AA Members' Handbook 1982/3*

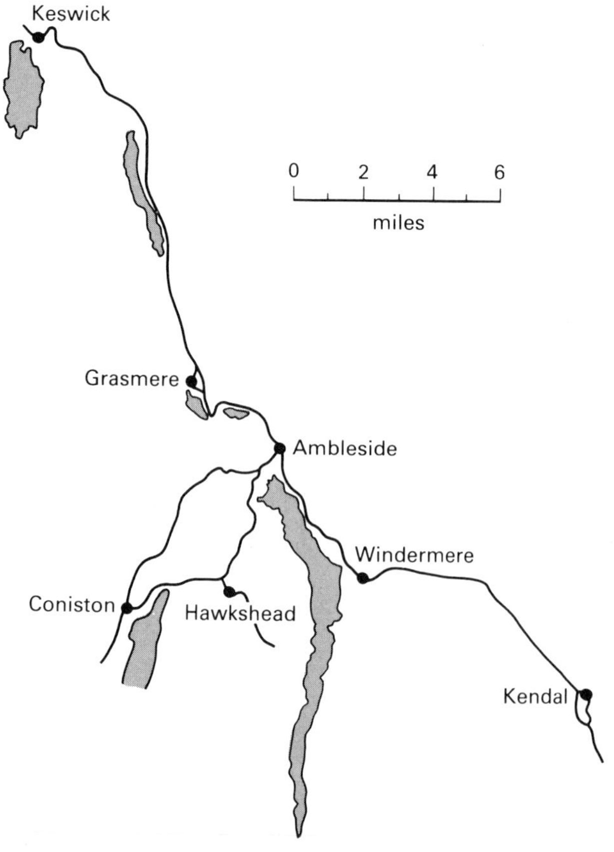

Processing data and presenting results

When testing several hypotheses it is best to process the data and present the results for each hypothesis together.

Hypothesis 1

The interaction statistic can be used to predict traffic flow between settlements.

1 Calculate the interaction statistic. This provides an index of the predicted movement between two settlements. Use the following formula:

$$M_{ij} = \frac{P_i \times P_j}{(d_{ij})^2} \times 100$$

where P_i = size of settlement i
P_j = size of settlement j
d_{ij} = distance between settlements i and j

or

$$\text{Interaction statistic} = \frac{\text{Population town A} \times \text{population town B}}{(\text{distance between A and B})^2} \times 100$$

2 Tabulate the results of your traffic count (fig. 10.3).

3 Present the results of the traffic count and the interaction statistic so that they can be compared. A flow diagram where the width of arrow is proportional to flow is a suitable method (fig. 10.4). Since the interaction statistic is only an index, different scales may be used for the two sets of figures. It might be better to draw two separate flow diagrams rather than cramming all the information

10.3 Traffic count taken at Ambleside, 7 May 1982 from 11.15 to 12.15

Road to	No of vehicles
Grasmere and Keswick	444
Hawkshead	182
Windermere and Kendal	456
Coniston	228

10.4 Flow diagram showing predicted and observed traffic flow

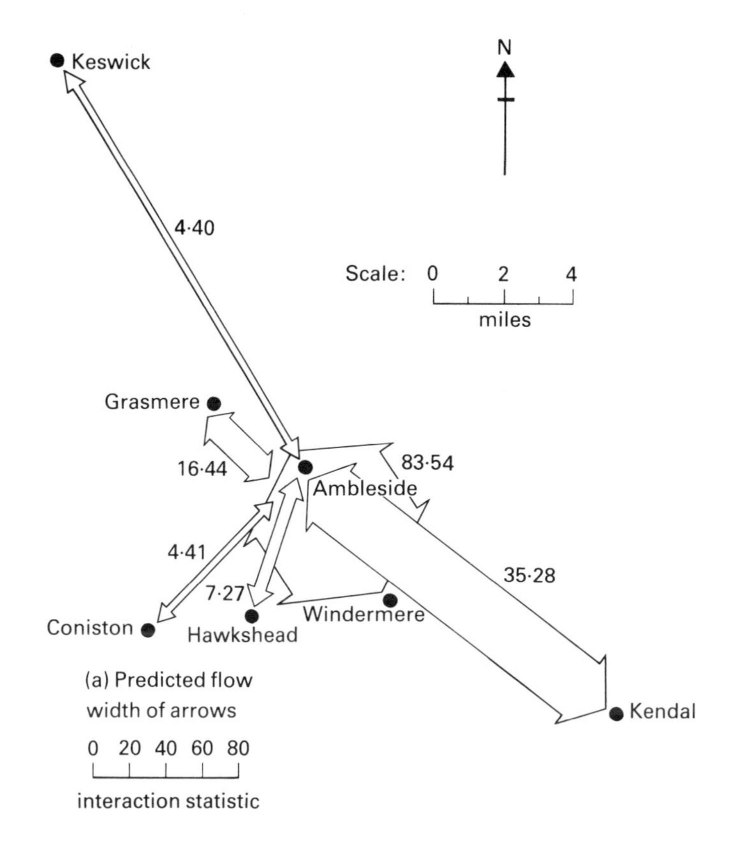

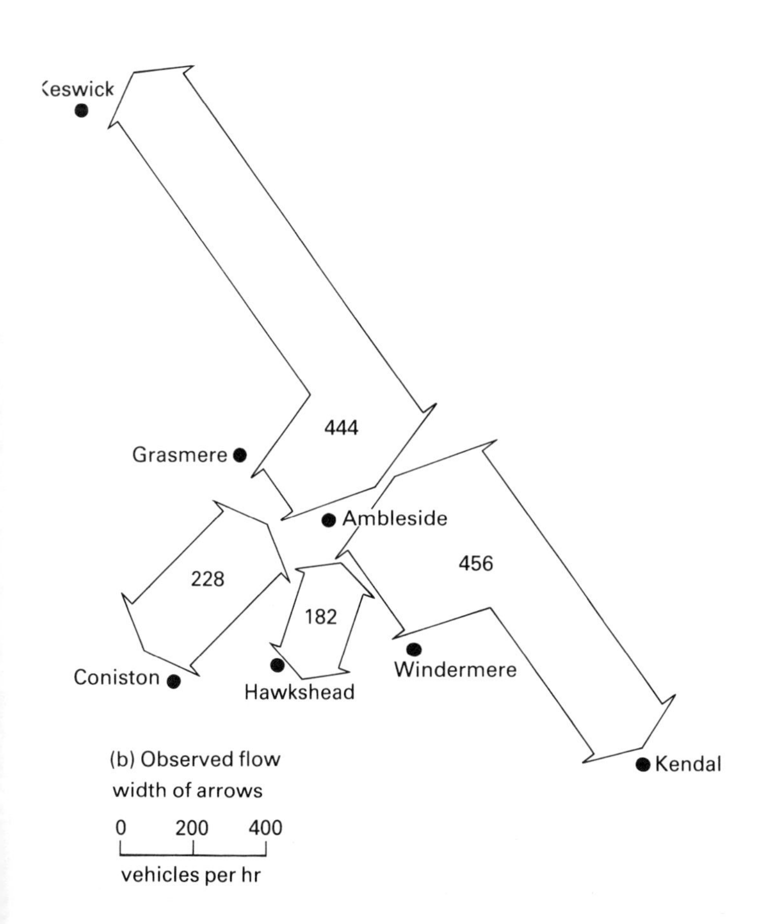

onto one diagram. You may have to show the predicted movement to two (or more) settlements in the same direction, because they are reached using the same road. This is illustrated in fig. 10.4.

Hypothesis 2

The bus service between settlements is related to the size of settlements served and the length of journey.

1 Tabulate your information about bus services to selected settlements (10.5).
2 Present this by means of a routed flow map (fig. 10.6). This shows the actual route between the settlements. Lines are constructed so that their width is proportional to the volume of flow at every point along the route. Be careful to use a scale that allows you to show all your data.

10.5 Buses from Ambleside

Settlement	Number of buses per week in summer	Interaction statistic
Grassmere	67	16.44
Hawkshead	53	7.27
Windermere	200	83.54
Kendal	69	35.28
Keswick	52	4.40
Coniston	51	4.41

10.6 Routed flow map showing number of buses from Ambleside per week in summer

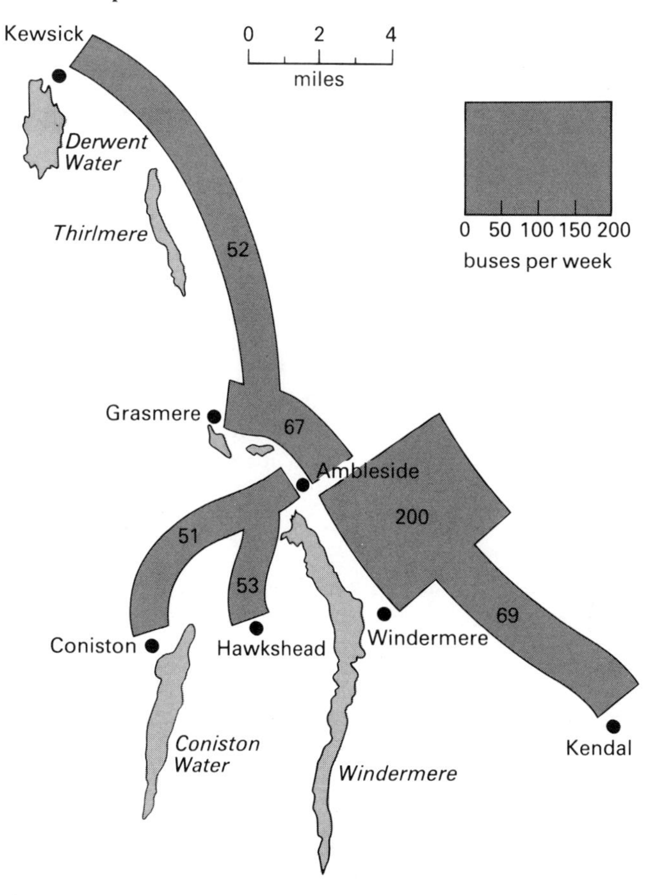

3 By comparing the number of buses on each route with the interaction statistic it is possible to look for a relationship between bus services, size of settlements served and the length of journey. A scatter graph should be constructed (fig. 10.7) to show this. The example given is on log–normal graph paper because of the large range in numbers of buses per week. The data could then be tested using Spearman's rank correlation if the scatter graph indicates a correlation (see chapter 2). (Spearman's rank correlation would not be appropriate for the data of fig. 10.7 as there are too few pairs.)

10.7 Scatter graph showing the relationship between the interaction statistic and the number of buses per week

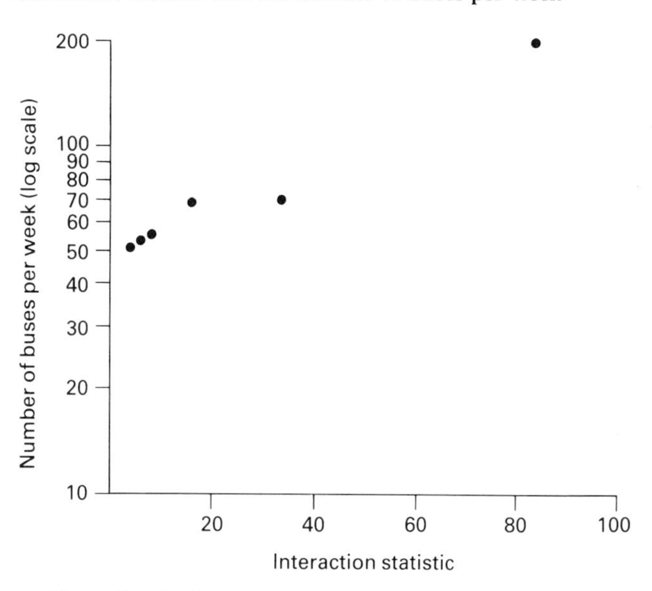

Hypothesis 3

Natural features influence the time taken to travel between settlements.

1 Use an OS map of the area to draw a simple topographical map to show the natural features of the area. You should show the main settlements in the area as well as all the smaller settlements for which you have collected information about time of journeys. You should also show the main features of relief and the rivers, lakes and coastlines.
2 Construct an *isochrone* map to show the time taken to travel by bus from the central town. This is an isopleth map where the lines join places of equal time. First draw a base map showing all the settlements for which you have obtained data. The base map must be the same scale as the topographical map but should show *only* settlements. Alternatively the map could be drawn onto a tracing paper overlay.

Plot all the times you have recorded in minutes. These are the control points and there must be at least 30, spread throughout the area. Decide on the number of isochrones to be drawn by looking at your data. It may be suitable to choose 5 minute or 10 minute intervals. As a guide (not a rule) the number of isochrones should not be more than 5 × the log of the number of observations.
For 30 observations: $5 \times \log 30 = 5 \times 1.4771 = 7.3$
Thus no more than 7 isochrones should be drawn (fig. 10.8).

Hypothesis 4

Data for your hypothesis concerning cost of travelling must also be processed and presented. The information could also be drawn on an isopleth map, in this case these are known as *isophore* maps.

Analysis

Your hypotheses must be discussed separately and conclusions drawn in each case. You should first describe your results and then look for some explanations. Remember that information obtained from OS maps or a road atlas may help your explanation, in addition to your own knowledge of the area. For example a lack of similarity between the observed and the predicted traffic flow from the interaction statistic could be due to tourist attractions or to the location of a factory or quarry away from a centre of population.

It is essential to remember that some of the data collected for your transport study may not be a true reflection of traffic and transport. Counting traffic for one hour on one day in the year gives a very small sample, therefore the information must be treated with care.

A general discussion of the public and private traffic flows is also needed to conclude your work. In this you must attempt to link together the different types of data collected and the various hypotheses tested. Make sure your work is presented as a whole and not as a series of seemingly unconnected sections.

Suggestions for further study

1 Look at the changes in traffic flow during a 12-hour (or 24-hour) period e.g. 07.00 to 19.00 and compare this with the typical pattern of traffic flow with its two daytime peaks. Traffic should be counted at the same place for 10 minutes each hour, and then presented as a percentage of the total traffic flow.
2 In a tourist area it may be possible to count the number of foreign cars as a percentage of the total number of cars using different routes.

Further reading

K Briggs, *Introducing Towns and Cities*, Hodder & Stoughton 1974
K Briggs, *Introducing Transportation Networks*, Hodder & Stoughton 1972
F Clegg, *Simple Statistics*, Cambridge University Press 1982
P Davis, *Data description and presentation*, OUP 1974
P Toyne and P Newby, *Techniques in Human Geography*, Macmillan 1971

10.8 Isochrome map showing times of bus journeys from Ambleside

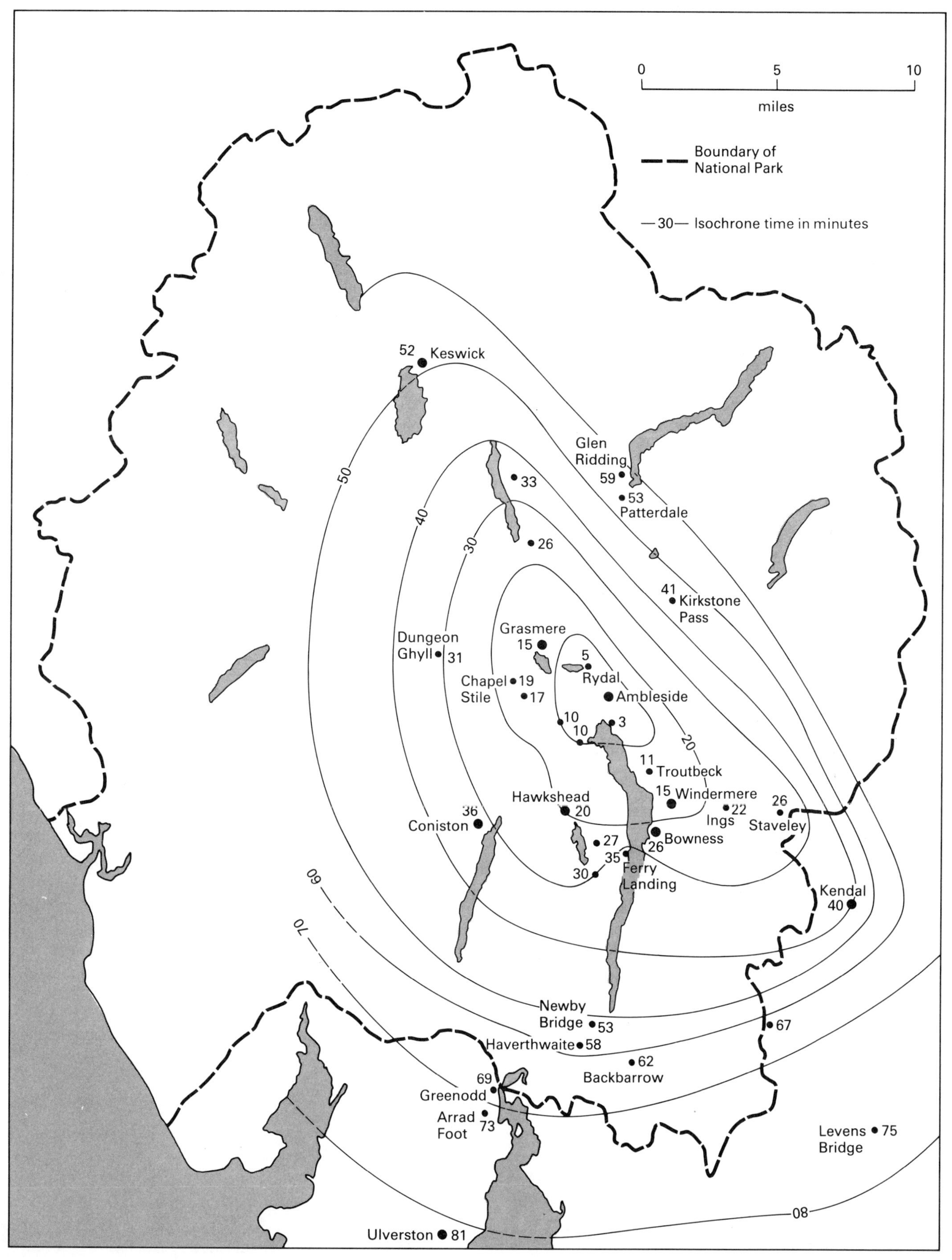

11 Hillslopes in carboniferous limestone

Suitable for individual or group study

Topic for study

The form of a hillslope is the result of the interaction of structure, process and time. This study aims to measure the form of hillslopes on limestone, a rock with readily recognisable structure, and to begin an investigation of the effects of process and time on this structure.

Hillslopes may be divided up into elements, or sections, of uniform gradients, separated by breaks of slope (points at which the angle of slope changes). Some hillslopes have been found to consist of waxing slope, free face, debris slope and waning slope, whereas others have a rectilinear slope between the waxing and waning slope (fig. 11.1). Ungraded slopes, which consist of several free faces separated by debris slopes, are common in some areas. For more detail about slope form consult Young, *Slopes*.

Processes operating on slopes include weathering and mass movement. Some of the elements observed today are the product of past processes of weathering and mass movement. Today in Britain freeze–thaw is of negligible importance but during glacial and periglacial times it was responsible for the production of scree. Solifluction was also a very important mass movement process but it is no longer important. Both processes were very effective in limestone areas, particularly on south-facing slopes which experienced large diurnal temperature ranges (see Hanwell & Newson, *Techniques in Physical Geography*).

Possible hypotheses

1 Waxing slope, free face (scar), debris slope (scree) and waning slope are readily identifiable slope elements in carboniferous limestone areas.
2 All free faces, scree slopes, waxing and waning slopes have developed their own characteristic angles.
3 Asymmetrical valleys develop where there are north and south facing slopes.
4 Scree slopes consist of material of differing sizes; there is a relationship between particle size and distance from the free face.
5 Screes are no longer actively forming and are becoming vegetated.

11.1 Elements of hillslopes

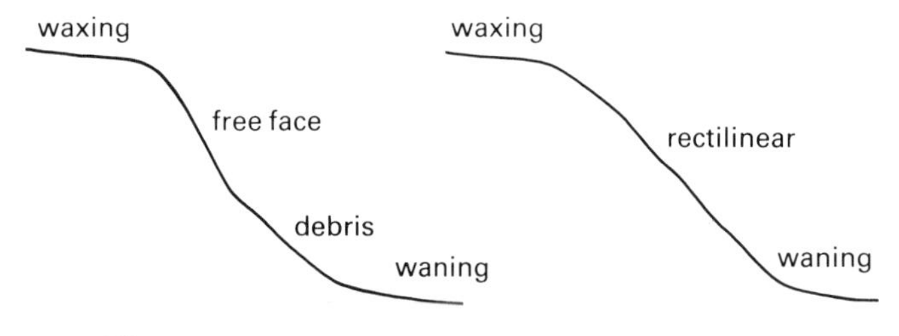

Location

The data collected in this study came from Lathkill Dale, Derbyshire but any W–E orientated limestone valley (with north and south facing slopes) is suitable. Do not choose a valley with dangerously steep sides.

Data collection

Equipment

30-metre tape
clinometer
2 ranging rods if working singly
compass (for preliminary survey)
quadrat
ruler or calipers
camera

Choice of site(s)

Make a preliminary survey of the valley and decide which areas are accessible for a slope transect. If you intend

11.2 Valley recording sheet

Transect no 1 from south to north facing slope

Reading no	Distance between breaks of slope (m)	Foresight (°)	Backsight (°)	Mean angle
1	8.9	15	17	−16
2	21.2	21	20	−20.5
3	6.2	7	9	−8
4	14.5	11	12	−11.5
5 Valley floor	40.6	0	0	0
6	5.7	15	16	+15.5
7	1.6	57	57	+57
8	1.7	35	30	+32.5
9	1.6	90	88	+89.0
10	9.9	19	21	+20.0
11	3.3	11	11	+11.0
12	2.5	25	24	+24.5
13	36.0	15	16	+15.5

11.3 A single transect

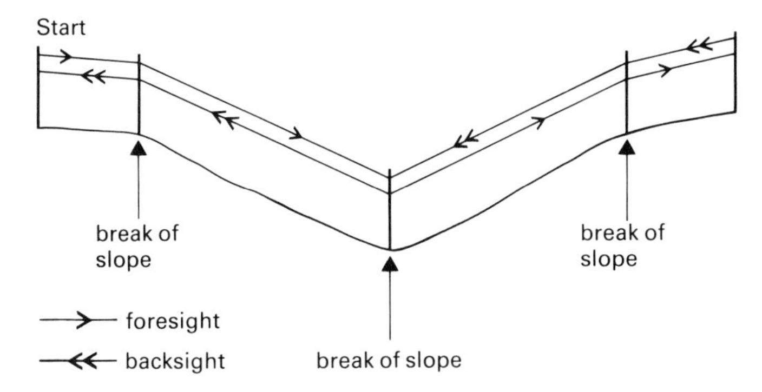

to test hypothesis 3 check that the orientation of the valley is W–E using the compass.

For hypotheses 1, 2 and 3 make at least 5 (and not more than 10) transects of the valley. Ideally they should be regularly spaced. Prepare sufficient recording sheets (fig. 11.2).

1 Start at the top of the waxing slope on the south facing slope. Stretch the tape downslope to the first break of slope and record the distance (length of slope facet) in metres. .
2 Use the clinometer to read the foresight and backsight. Use a partner the same height as yourself and align the clinometer with his/her eyes. If working alone insert ranging rods at the breaks of slope.
3 Calculate the mean angle and indicate whether you are working downslope (−) or upslope (+).

For Hypothesis 4, identify a free face (scar) with a debris slope (scree) about 25–50 metres in length.

1 Stretch the tape at right angles to the free face down the scree.
2 Using the quadrat make a regular sample of scree size downslope as shown in fig. 11.4. Measure the long axis (*l*) and intermediate axis (*i*) of the piece of scree at each intersection of the quadrat and record as shown (fig. 11.5).

11.4 Scree sampling

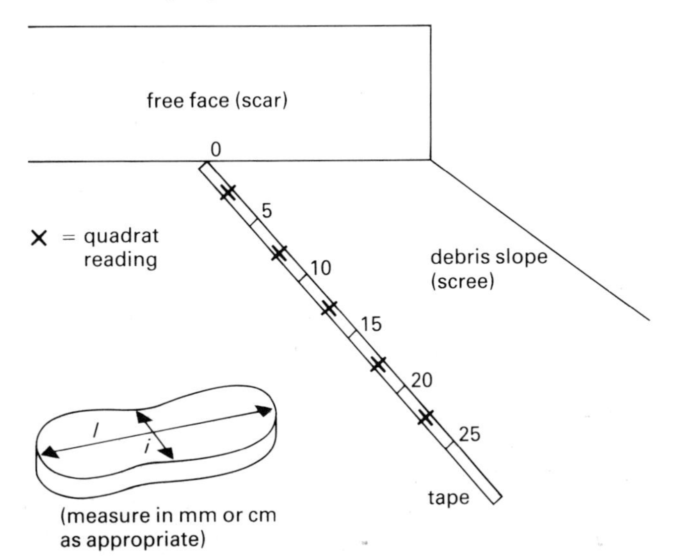

11.5 Scree recording sheet

Distance from free face (metres)									
0–4.9		5–9.9		10–14.9		15–19.9		20–24.9	
l	*i*	*l*	*i*	*l*	*i*	*l*	*i*	*l*	*i*
10	5								
10	6								
9	4								
etc.									

For hypothesis 5, identify several free faces (scars) with debris slopes (screes) beneath them. They need not be accessible. Photograph the screes from the front elevation (not at a side elevation angle). The photograph will need to be enlarged for analysis.

Presentation of results

1 Add together the total distance for each transect and select a suitable scale. For example, if longest transect is 150 m a scale 4 mm = 1 metre would be suitable.
2 Draw out transect 1, beginning in the centre of the graph paper at the left-hand side.
3 Now superimpose the other transects over the first, beginning at the same point. When all transects have been inserted, in pencil, go over each one in a different colour.
4 Mark breaks of slope and label each element as shown in fig. 11.6. Mark the north and south facing slopes.
5 Draw a dispersal diagram to show the range of angles measured on each part of the slope (waxing, free face, debris, waning, rectilinear). See fig. 11.7. Note that some sections (elements) of the slope will be made up of several facets.

   Use the results from all transects, plotting dots in a different colour for each transect.
6 Calculate the *mean* angle for each element, using data from all transects. Plot your results on a bar graph *in descending order of steepness* – see fig. 11.8 but do not assume that your data will follow the same descending order.
7 Draw a bar graph to show the number of free faces and the number of rectilinear slopes on north- and south-facing

11.6 Drawing transects (slope elements have been labelled for transect 2 only)

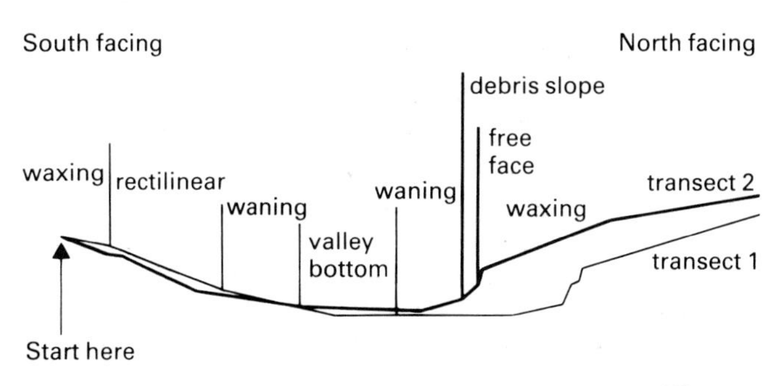

11.7 Data for transect 2

Reading no	Distance (m)	Mean angle (°)	Slope element
1	1.8	25.5	rectilinear slope
2	2.8	27.5	
3	3.5	21	
4	5.5	15.5	
5	2.5	18	
6	9.5	22	
7	6.7	10.5	waning slope
8	8.4	9.0	
9	11.2	2.5	
10	12.2	0	valley bottom
11	5.7	5.0	waning slope
12	5.0	15.25	
13	1.7	40.5	debris slope
14	2.5	72.5	free face
15	2.4	12.5	waxing slope
16	3.3	23.5	
17	9.3	19	
18	30.6	7.5	

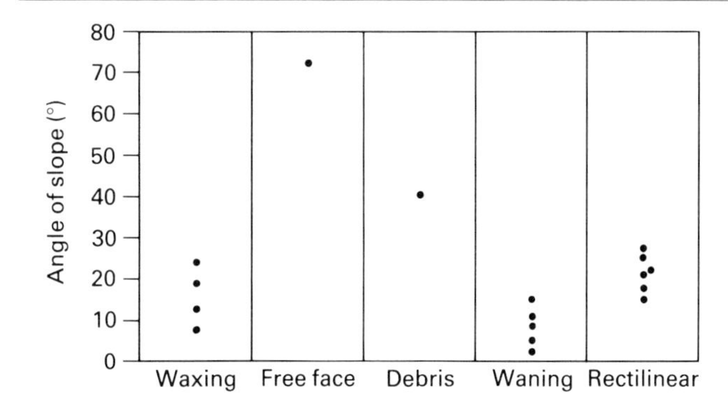

11.8 Bar graph showing mean angle

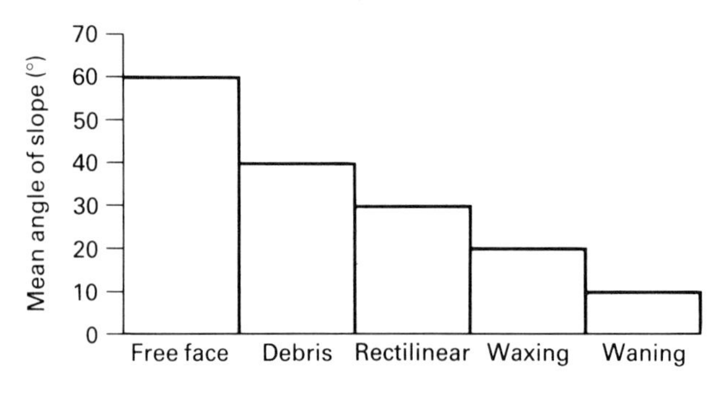

slopes (see fig. 11.9). Use data from all your transects.

8 Investigate the relationship between length of facet and slope angle for all data by drawing a scatter graph. Put the length of facet on the x axis. Mark north and south facing facets with different coloured dots. Try to draw in the best fit line. If your graph is inconclusive test the data by using Spearman's rank correlation (see chapter 2) though remember that

(*a*) you must not have more than 30 pairs of data,

11.9 Bar graph showing numbers of faces

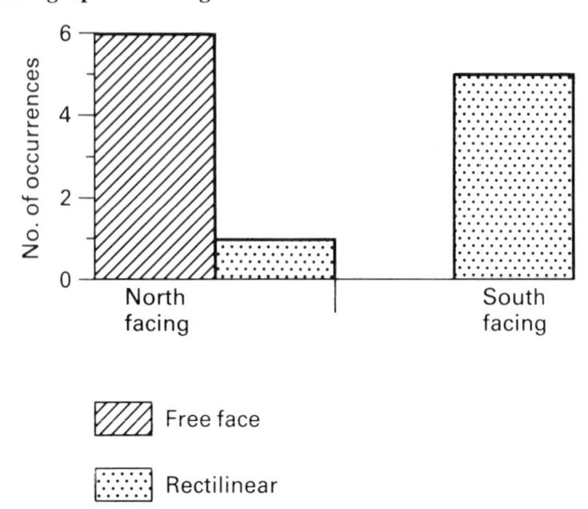

	North facing	South facing
Free faces	6	0
Rectilinear slopes	1	5

11.10 Contingency table showing size index at different distances from the free face

Dist. from free face (m)	Size index 0–50	51–100	101–150	151–200	Total
0–4.9	15	5	3	2	25
5–9.9	12	7	4	2	25
10–14.9	7	9	6	3	25
15–19.9	3	11	6	5	25
20–24.9	2	2	8	13	25
Total	39	34	27	25	125

(*b*) Spearman's test will not distinguish between north and south facing slopes.

Remember to set a hypothesis before you use the test or draw the graph.

9 Calculate the *size index* for each scree particle ($l \times i$) Set out your results in a contingency table (fig. 11.10) Using chi-square test the relationship between particle size and distance from the free face. Remember to set a null hypothesis H_0 (reword the second part of hypothesis 4). Before you start check that your data is appropriate for this test (see chapter 12).

Alternatively

10 Draw a histogram to show the variations in size index throughout the whole scree. Use the *column totals* from the contingency table, and subdivide each column to show distance from the scree as in fig. 11.11.

11 Place a sheet of tracing paper over each scree photograph. Divide the paper into 10 squares (or a multiple of 10 depending on the size of the photo). Estimate the percentage of bare rock in each square. Write in the

11.11 Size index histogram

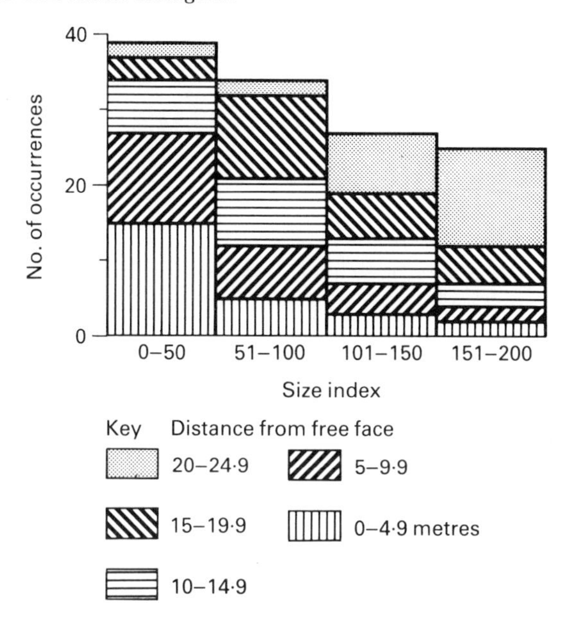

11.12 Estimated percentages of bare rock on scree

45	50	55	60	55
55	75	75	80	60
50	70	100	80	65
50	70	70	75	55
40	45	45	50	45

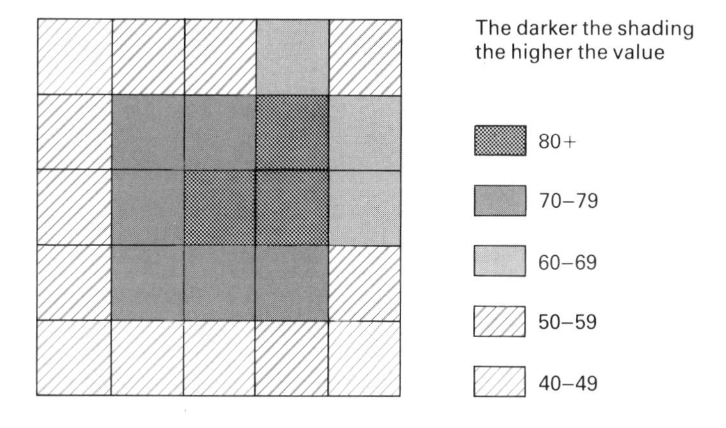

estimated percentages lightly in pencil.

Use the choropleth technique to shade your tracings in different colours or density of shading, to show the change in quantity of bare rock. Alternatively draw out the grids onto plain paper before shading them.

Analysis

Visual analysis of the slope transects will show whether or not limestone slopes contain the four slope elements mentioned in hypothesis 1 and will also reveal asymmetry of present (hypothesis 3).

The dispersal diagram (11.7) will indicate whether similar slope elements on different transects display comparable angles. Should hypothesis 2 be proved you should consider whether similarity of angle can be related to the structure of the rock and/or to processes operating on the slope. Limestone, because of its close jointing and permeability exerts structural control over weathering and mass movement processes. (You could investigate these in 'further study'). The mean angle bar graph (11.8) summarises your findings for hypothesis 2.

The number of free faces and rectilinear slopes on north and south facing slopes, if significantly different, indicate that different processes were operating on these two slopes at one time. Try to explain why south facing slopes have more rectilinear facets should you obtain results similar to those in the example given (fig. 11.9).

The chi-square test and/or histogram will provide an answer to hypothesis 4. Consider why different size scree particles exist and why there appears to be some sorting downslope if your results produce these relationships. If your photograph analysis reveals that the proportion of bare rock on the screes is low, you can conclude that the screes are fossil features and are not being formed at present. However, some mass movement is still possibly occurring on the screes, lowering the angle.

Further study

1 If you have found that long slope facets have shallower angles try to explain this in terms of slope evolution.
2 Compare slopes on limestone with those on another rock type.
3 If you have found that slope angles on different elements are similar relate this to theories of slope evolution (see Young, *Slopes*). Consider the circumstances in which weathering and mass movement *maintain* the slope angles of each element and discuss whether such circumstances are likely to be prevalent today in limestone areas.
4 Compare a valley with west and east facing slopes with the one you have first studied to ascertain whether or not it is asymmetrical.

Further reading

M Clark and R J Small, *Slopes and Weathering*, Cambridge University Press 1982
K C Edwards, *The Peak District*, Collins 1964
R K Greswell, *Physical Geography*, Longman 1967
J D Hanwell and M D Newson, *Techniques in Physical Geography*, Macmillan 1973
B J Knapp, *Practical Foundations of Physical Geography*, Allen & Unwin 1981
A Young, *Slopes*, Longman 1975

12 Location of industry in towns

Suitable for individual or group study

Topic for study

All towns possess some industries. In some cases the industry is concentrated into one or two clearly defined areas such as along a river valley, canal or railway line or along a coastline. These industrial areas tend to dominate discussion of industrial location within towns. However most towns also possess a wide variety of industry which is scattered throughout the town.

It is usually possible to identify four different groups of industry within a town each of which can be found in characteristic locations.

1 *Centrally located industries* are usually small scale industries operating from small workshops close to the Central Business District. Many will have been established during the nineteenth century and have survived in their original locations due to inertia, e.g. gunsmiths in Birmingham or cutlers in Sheffield. Some of these industries may be directly linked to a retail outlet and thus need a central location, e.g. silversmiths and jewellers.
2 *Heavy industries*, such as iron and steel, textiles and electricity generation, rely on heavy or bulky raw materials. These industries need large areas of land and cheap transport and therefore usually group together next to rivers, canals or railway lines or along a coastline, often forming a clearly defined industrial sector.
3 *Light industries* use small and light raw materials or components and therefore tend to rely on road transport. These industries have mainly been established in the second half of the twentieth century often in fairly small premises. This means that light industry is often located in suburban areas close to main roads or motorways where land is available and not as expensive as in central areas. These industries can often be found on purpose-built industrial estates on the outskirts of towns but are sometimes found in disused factories such as cotton mills which have been adapted for light industrial use. Examples of light industries include electronics, light engineering and canned food.
4 *Market oriented industries* may be found in any part of the city from the CBD to the outer suburbs. These industries rely on direct contact with their customers, either because they provide a personal service, such as made-to-measure tailoring or hand-made furniture, or because their products are perishable, such as ice cream or bread and cakes. These industries are unusual in that they are likely to benefit more from a location at some distance from their competitors than from close proximity and linkage.

Interesting fieldwork can be carried out studying locations of different types of manufacturing industry within a town to see if these groups of industry exist and whether their typical locations can be identified.

Possible hypothesis

Old established industry tends to be centrally located whilst food processing industries are scattered throughout suburban areas.

Location

Choose a large town (over 100,000 population) which has a wide variety of industry. It must be a town which is covered by *Yellow Pages* and one for which you can obtain a comprehensive street guide such as the *A to Z* and a clear large scale street map.

You do not need to be familiar with the town, although some knowledge will make the study easier, because the investigation çan be based entirely on secondary sources, but you do need to have access to information about the town's industry. The data for this example are taken from both Leicester and Sheffield.

Data collection

In order to collect your data you will need the following resources for the town you have chosen:

Yellow Pages
Postcodes book, published by the Post Office
Large scale street map
Street guide such as *A to Z*

The *Yellow Pages* and *Postcodes* book are available in most public libraries for reference purposes.

1 Use the *Postcodes* book to establish the position of the postal districts in the town you have chosen. A map is provided in the front of the book. You will probably find that some of the postal districts are outside the city boundary and you may wish to omit them from·your study. For example, in Leicester, only districts LE1 to LE5 lie within the city boundary, districts LE6 to LE17 include surrounding settlements such as Market Harborough, Lutterworth and Loughborough.
2 Find out what the old established industries are in the town you have chosen. You may find a regional geography

12.1 Location of selected manufacturing industry in Leicester

Manufacturing industry	LE1	LE2	LE3	LE4	LE5	LE6	LE7	LE8	LE9	Totals
OLD ESTABLISHED INDUSTRY										
Tapes, webbings & binding			1	1	1	1				4
Handbags	2	1	2	1	1					7
Braces, belts & suspenders	3	2	1							6
Button & buckle coverers	2	2	1							5
Children's wear		2		5	2			1		10
Dyes		1	1				1			3
Badges & emblems	1									1
Dyers & finishers	1	5	4	3				1		14
Totals	9	13	10	10	4	1	1	2	0	50
FOOD PROCESSING INDUSTRY										
Bacon & ham curers								1		1
Bakers	1			1			1	1	1	5
Biscuits		1						1		2
Confectionery			1	2	1		1			5
Sausage makers								1		1
Pie makers	1			1			1	2		5
Other food products	1			2	1		1		1	6
Soft drinks				2				1		3
Ice cream	1	1	2		2	1				7
Dairy products		1	2	1	1			2		7
Totals	4	3	5	9	5	1	4	9	2	42

N.B. The data should include information for at least 40 manufacturers in each group and preferably for many more than this.

textbook useful. It is essential that you identify these industries correctly so ask for advice at this stage if necessary. Go through the *Yellow Pages* and extract information about old established industries in the town. Record the address of each industrial concern. Remember that the entries will be in a number of different classifications. For example, in Leicester, old established industries include braces, belts and suspender manufacturers, dye manufacturers, button and buckle cover manufacturers etc.

3 Use the *Postcode* book to find the postal area of each address and record this. For example,
Badge and emblem manufacturers, Wellington Street, LE1
You should collect as much data as possible. Some classifications may have over 100 entries each and ideally you should record each one, alternatively you could *sample* your data (see Appendix A).

4 Repeat this procedure for food processing industries.

Processing data

Construct a matrix to show the distribution of the industries according to their postal districts (fig. 12.1).

Presentation of results

1 Construct proportional circles to show the distribution of old established industries in each postal district. To calculate the radius of the circle first find the square root of the number of industries in each district then work out a suitable scale.
Construct the circles on a base map showing the postal districts. You may find that circles overlap particularly if the base map has a small scale. However, this can serve to emphasise concentrations (fig. 12.3).

12.2 Old established industry in Leicester

	No of industries	Sq. root	Length of radius
LE1	9	3	12 mm
LE2	13	3.6	14.4 mm
LE3	10	3.16	12.6 mm
LE4	10	3.16	12.6 mm
LE5	4	2	8 mm
LE6	1	1	4 mm
LE7	1	1	4 mm
LE8	2	1.41	5.6 mm

2 Repeat the procedure for food processing industries using a second base map or a tracing overlay. Remember to use a map of the same scale and to calculate length of radius in the same way so that you can compare the results.

3 Select industries which are well represented in your data e.g. dyers and finishers, and ice cream manufacturers.

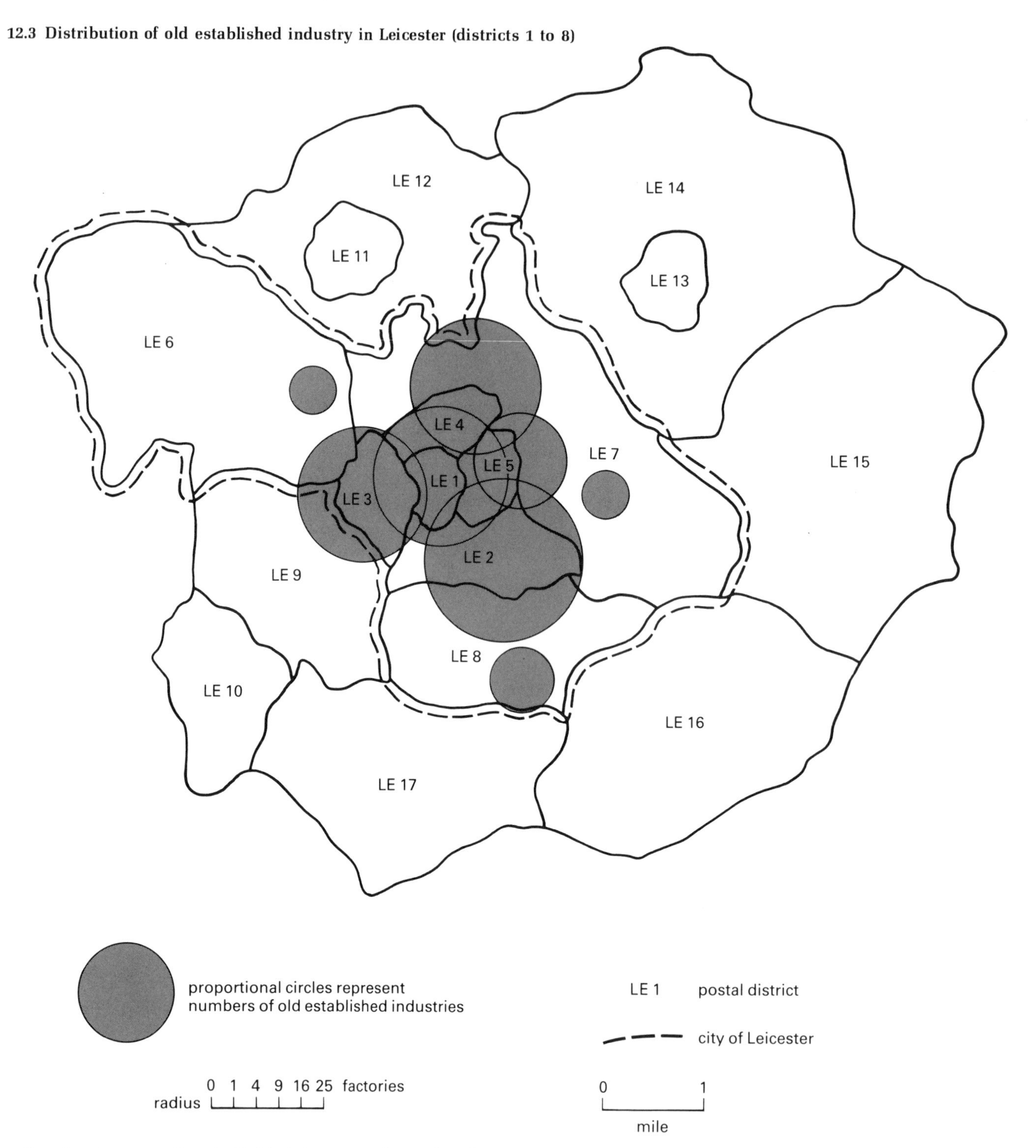

12.3 Distribution of old established industry in Leicester (districts 1 to 8)

Construct proportional divided bars to show the distribution of these particular industries within the town (fig. 12.4).

This information could be converted to percentages and presented on bars of equal length but this could be misleading as it would hide the larger numbers of some industries.

The predominance of the dyers and finishers and children's wear manufacturers in postal districts LE2, LE3 and LE4 is clear. Reference to a map of the town may suggest reasons for this. Similarly the widespread occurrence of dairy products and ice cream manufacturers can be seen.

There are often large numbers of entries for particular industries in a town and the more time you spend collecting and recording these then the more reliable your results and conclusions will be.

4 Choose one or more industries which appear to have a typical location. Map the location of the selected industries on a large scale street map using the addresses you initially recorded.
5 Transfer your results onto a smaller scale base map if possible (see fig. 12.5). The transfer of information from a large scale street map to a smaller scale map will

12.4 Proportional divided bars showing the distribution of four industries in Leicester

Dyers and finishers

Children's wear

Ice cream manufacturers

Dairy product manufacturers

0 1 2 factories

LE 1 LE 2 LE 3 LE 4

LE 5 LE 6 LE 8

obviously lead to some approximation and loss of detail, however it may bring out some interesting patterns within the town as a whole. Fig. 12.5 shows this information for cutlers and silversmiths in Sheffield.

6 The chi-square test is a useful statistical test for looking at distributions and it is possible to use it for data of this type. You *must* check your own data against the requirements of the test before using it.

1 There must be at least 20 observations
2 The data must be in the form of frequencies. Percentages must *not* be used.
3 No expected frequency must be less than 1 and no more than 20% of expected frequencies below 5 (N.B. categories can be combined).
4 The observations must be independent.

$$\chi^2 = \sum \frac{(O-E)^2}{E}$$

where O = observed frequency
E = expected frequency

Null hypothesis = There is no difference in distribution of old established industries and food processing industries in Leicester.

12.5 Sheffield postal districts: location of cutlers and silversmiths

12.6 Observed frequency of Leicester industry

	LE1	LE2	LE3	LE4	LE5	LE6 to LE9	Row total
Old established industry	9	13	10	10	4	4	50
Food processing industry	4	3	5	9	5	16	42
Column totals	13	16	15	19	9	20	92

$$\text{Expected frequency} = \frac{\text{row total} \times \text{column total}}{\text{grand total}}$$

12.7 Expected frequency of Leicester industry

	LE1	LE2	LE3	LE4	LE5	LE6 to LE9
Old established industry	7.06	8.69	8.15	10.33	4.89	10.87
Food processing industry	5.93	7.30	6.85	8.67	4.11	9.13

Only 2 out of 12 (16.66%) expected frequencies are below 5, therefore we may proceed with the test.

12.8 $(O-E)^2 \div E$ for Leicester industry

	LE1	LE2	LE3	LE4	LE5	LE6 to LE9
Old established industry	0.53	2.14	0.42	0.01	0.16	4.34
Food processing industry	0.63	2.53	0.50	0.01	0.19	5.17

$$X^2 = \frac{(O-E)^2}{E} = 16.63 \quad df = (r-1)(c-1) = 5$$

On consulting the significance tables for χ^2 the result shows that we can reject the null hypothesis (H_0) at the 0.01 level, that is very confidently, because the χ^2 value obtained is higher than the value shown in the table. In other words the distribution of old established industry and food processing industries in Leicester is significantly different.

Analysis

The results should lead to interesting discussion of industrial location within the town, although you may find that the locations of some groups of industries do not fit into the expected pattern whilst others do.

As usual it is not adequate simply to describe the results you have found and to reject or accept the hypothesis. You must look for explanations of the patterns and suggest reasons for your results. In the case of old established industries you will need to look at why they are centrally located and if there are exceptions to this pattern, you must discuss possible reasons for their location elsewhere. Look for the influence of physical features on their location or for evidence of Government influence such as development of industrial estates.

Similarly the location of food processing industries will need both description and explanation. One thing to remember is that some food processing industries will need to be near their customers whilst others are found on industrial estates. You should also try to explain why some areas of towns have very little industry whilst others have considerable concentrations.

Suggestions for further study

1 Use the location quotient to look at the distribution of industries within different areas of the town (Mowforth, *Statistics for Geographers*, p. 49).
2 The *Yellow Pages* provide vast amounts of data which can be used for locational exercises. In addition to studying different groups of manufacturing industry it would also be possible to study location of offices, perhaps looking at location of
   (*a*) professional offices (e.g. solicitors, accountants, surveyors),
   (*b*) head offices of manufacturing industry,
   (*c*) administrative offices, particularly local government,
   (*d*) banks and building societies.

Further reading

J Bale, *The Location of Manufacturing Industry*, Oliver & Boyd 1976
P McCullagh, *Data Use & Interpretation*, OUP 1974 (chi-square test pp. 6–12)
M Mowforth, *Statistics for Geographers*, Harrap 1979 (chi-square test p. 30)

13 Suburbanised villages

Suitable for individual or group study

Topic for study

Since 1945 there has been careful planning of towns and cities in Great Britain, one aspect of which has been to control urban sprawl. In addition there has been increased affluence and car ownership during the last 30 years. One of the results of these changes has been the rapid expansion of villages within easy reach of large towns and cities. Some villages are entirely engulfed by the town but many are far enough away to remain as distinct places. These settlements are known as suburbanised villages or commuter villages or dormitory settlements.

Suburbanisation usually causes considerable change to both the morphology of the village and the socio-economic characteristics (e.g. age–sex structure; occupations) of the residents and this provides an interesting area of study. The model of the morphology of a suburbanised village, fig. 13.1, can form a useful framework of such a study.

13.1 The morphology of a suburbanised village

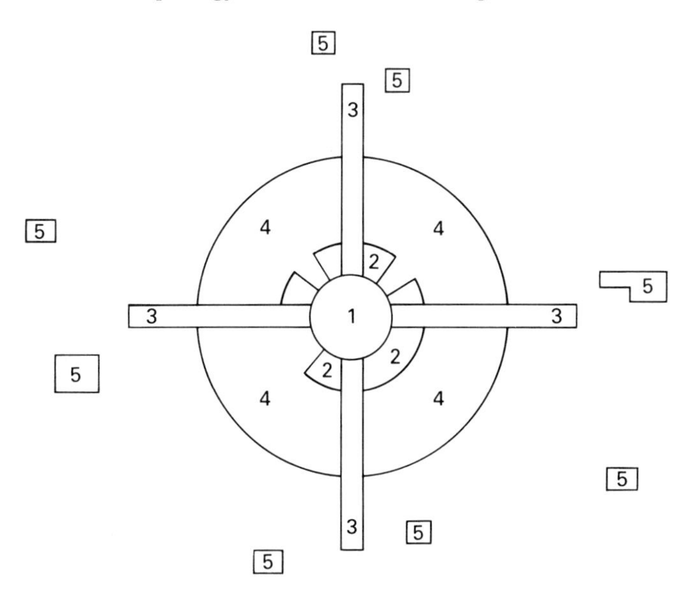

Zone 1: *Original village core* with infills and modifications
Zone 2: *Accretions*, small additions in close proximity to village core 1918–39
Zone 3: *Ribbon development*, 1920 to 1960
Zone 4: *Adjunts*, large modern estates, both council and private, mainly built since 1950
Zone 5: *Isolates* dispersed settlements around village but separate from it. Both old and modern buildings mainly farms and cottages

Possible hypotheses (for a named village you have chosen)

1 The chosen village reflects the model of a suburbanised village.
2 There are clear differences between the socio-economic characteristics of established residents and newcomers to the village.

Location

When you choose a village to study look for some evidence of suburbanisation, for example, a village where there has been recent new building and where the population has increased considerably. The census will provide information about population change but it is essential to visit any village you propose to study to establish whether there has been any new building.

The village will probably be within 15 miles of a large town and possibly much closer, it could even be a village now engulfed by the town. It should have a population of between 1,000 and 3,000 and should display the characteristics of a village not a small town in terms of the services it provides (see Whynne Hammond, *Elements in Human Geography* chapter 14).

Data collection

1 Obtain a large scale map of the village for example 1 : 10,000 or 1 : 2,500 OS maps. The map is needed to record information therefore it may be easier to draw your own base map for your fieldwork. Prepare several copies.
   A problem you may come across is that your map is not up to date and does not show recent buildings. If this is the case you must draw the extra buildings onto your map when you are in the field.
2 Walk round the village and classify each building according to the following system (it may be necessary to estimate some dates):
   (*a*) *Original village buildings* such as farms and cottages built any time before 1914.
   (*b*) *Accretions*, houses built in the inter-war period 1918–1939, mainly individual houses or short terraces.
   (*c*) *Ribbon development*, often semi-detached brick built housing built along roads between 1920 and 1960.
   (*d*) *Adjuncts*, large modern estates, council or privately built since 1950. You should indicate tenure on your map.

(e) *Infilling*, modern housing built since 1945, singly or in small groups not on housing estates.
(f) *Isolates* or dispersed settlement of any age.

Devise a key for ease of recording, and before you visit the village go to your local library to find out about typical building styles.

3 Record evidence of modifications to older housing, for example, house extensions or new garages. Use the same base map to record this information.
4 Visit the local Planning Office to collect information about building plans and policies for the village. You should certainly find out the dates of the main phases of building in the village.
5 Conduct a questionnaire in the village by visiting houses throughout the settlement. At least 50 questionnaires should be completed and preferably many more than this. You must try not to take a biased sample therefore you will need to visit all types of houses and try to question people of different age groups. You will also have to visit the village on different days of the week and at different times in an effort to include all groups of people.

A possible questionnaire is shown in fig. 13.2. This is only a guide but when devising your own questions you must think about how you will process the information you collect and whether or not it helps you to examine your hypothesis.

If you wish to visit people's houses to ask questions you must be prepared to explain the purpose of your visit clearly and politely. You should also take with you a letter from a teacher or other person in authority which explains the purpose of your questionnaire.

Processing and presenting data

1 Present a neat and clearly labelled land use map showing the classification of buildings and any recent modifications noted. Select colours to show different periods of growth. Draw a diagram to show the model of a suburbanised village using the same colours, so that a comparison can be made.
2 Analyse the questionnaire and tabulate all the results. You should try to divide your results into two groups, those relating to established residents and those for newcomers to the village. The criteria you use to distinguish these groups must be stated clearly.
3 Present the results of your questionnaire to show socio-economic characteristics of the established residents and the newcomers. Choose a variety of methods to show this

13.2

This questionnaire is only a guide. When devising your own you must choose time intervals which reflect the main phases of growth in the village you have selected. This information should be obtained from the local Planning Office.

1. How long have you been living in this village?
(select your own time intervals)
a) less than 5 years b) 5 to 10 years
c) 11 to 15 years d) over 15 years

If the answer to uqestion 1 is (d), over 15 years, then go on to question 3.

2. When did you move into the village?
Why did you move into the village? (More than one reason may be given)
a) liked the house b) wanted to live in the countryside
c) cheaper house price or rates d) other reason

3. What is the tenure of your house?
a) owner occupied b) council rented
c) privately rented d) other

4. How many cars do you have in your household?
a) none b) one
c) two or more

5. Where do you work?
a) in the village b) in the local area (record name of workplace?
c) in the main town(s) (record name of workplace) d) elsewhere (record name of workplace)

6. Where do you usually shop for most of your food?
a) in the village b) in the main town(s) (record names)
c) in both village and town d) elsewhere (record name of settlement)

7. Into which of the following age groups do you fall?
a) 18-30 years b) 31-45 years
c) 46-60 years d) over 60 years

13.3 Divided bars showing housing tenure

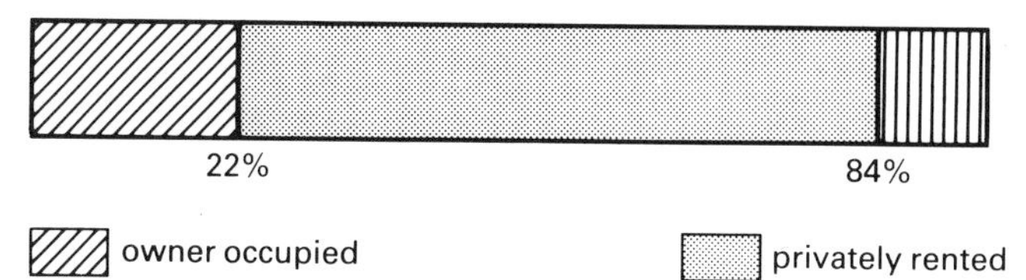

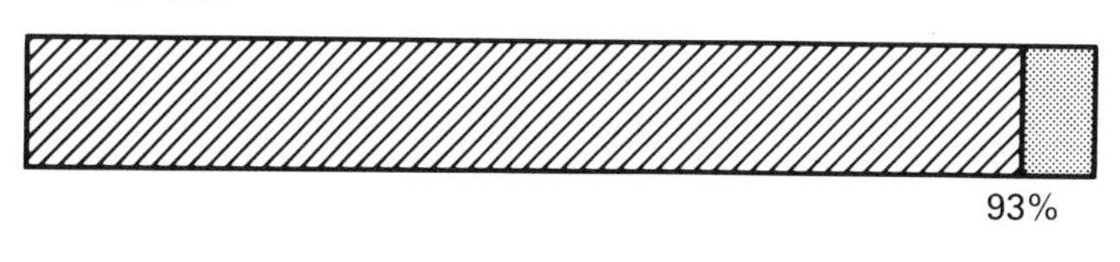

13.4 Histograms showing age distribution

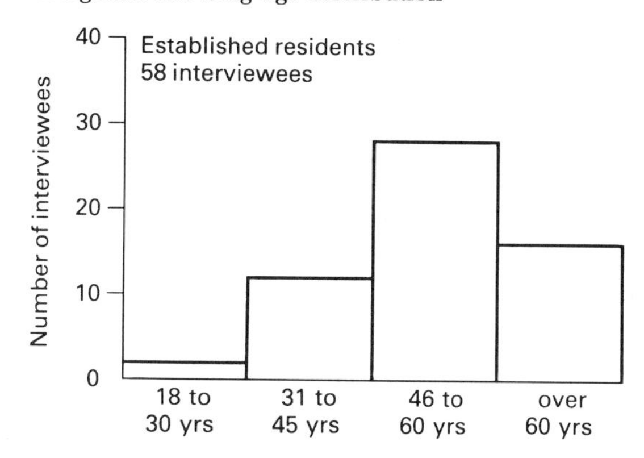

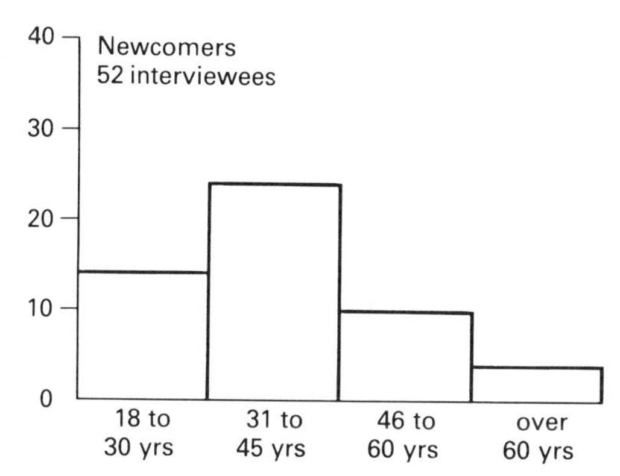

13.5 Pie charts showing car ownership

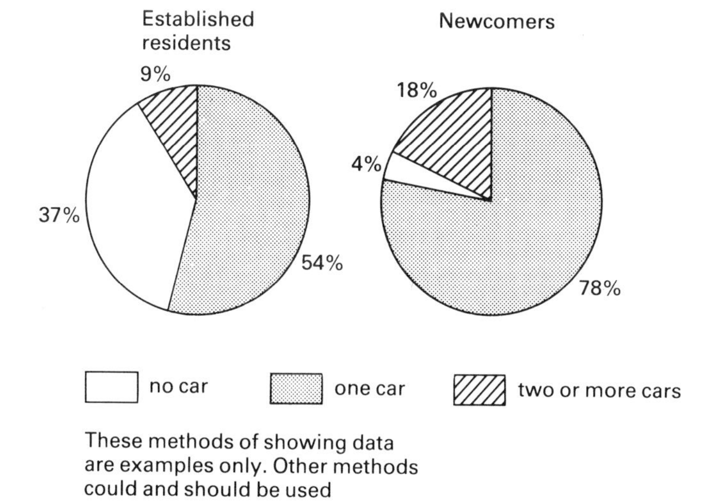

These methods of showing data are examples only. Other methods could and should be used

information. Some examples are given in figs. 13.3, 13.4 and 13.5 – others could be used. Think carefully about the methods you choose in relation to the information you are showing and try to include maps to show the spatial element of your work where possible. For example the information about workplace (question 5) should be shown on a map.

Analysis

There are two hypotheses involved in this study and you should discuss each of these separately. First look carefully at your map showing the morphology of the village you have studied and describe the growth of the settlement then discuss the probable reasons for growth. Your visit to the local Planning Office should provide information about this.

Then compare the morphology of the village with the model of a suburbanised village and describe the similarities and differences. No village will fit the model exactly but you should be able to suggest reasons for differences you observe. For example particular features of relief, or the activities of certain land owners or of the local council may have influenced village growth. From here you should be able to reject or accept your first hypothesis.

The second hypothesis considers differences in socio-economic characteristics of established residents and newcomers to the village. The data collected by questionnaire should be discussed and you should again give reasons for the patterns you observe. However you must remember the inherent problems of collecting information by questionnaire and the possibilities that your sample might be too small or biased. Your results should enable you to draw some conclusions, and thus you should be able to reject or accept your second hypothesis.

Suggestions for further study

1 Look at the Census figures for the village you have chosen and try to identify changes in patterns of the residents over time, for example 1951, 1961, 1971 and 1981.
2 Select a number of villages at varying distances from a large town and attempt to identify the importance of distance as a factor in causing suburbanisation. The use of data about house prices might be informative.
3 Compare the growth of villages within a National Park with villages which have less strict planning controls.

Further reading

W Hornby and M Jones, *An Introduction to Population Geography*, CUP 1980
C Whynne Hammond, *Elements of Human Geography*, Allen & Unwin 1979

14 A comparison of glacial end moraines

Suitable for individual or group study

Topic for study

In the mountains of North Wales, the Lake District and Scotland lie many fresh moraines dating from the very last Ice Advance (the Allerod Cold Phase, Zone III of the Late Glacial Period, 11,000 years ago). Although these moraines show little post glacial modification by weathering and soil creep it is noticeable that they become fresher and more hummocky as they are traced back from the valleys towards the hills. A succession of end (or recessional moraines) can be found, each marking an advance or standstill phase in the mass budget of the glacier. Fig. 14.1 shows how they develop. (For reference on chronological significance of moraines, see Embleton and King, *Glacial Geomorphology*.)

14.1 The formation of moraines

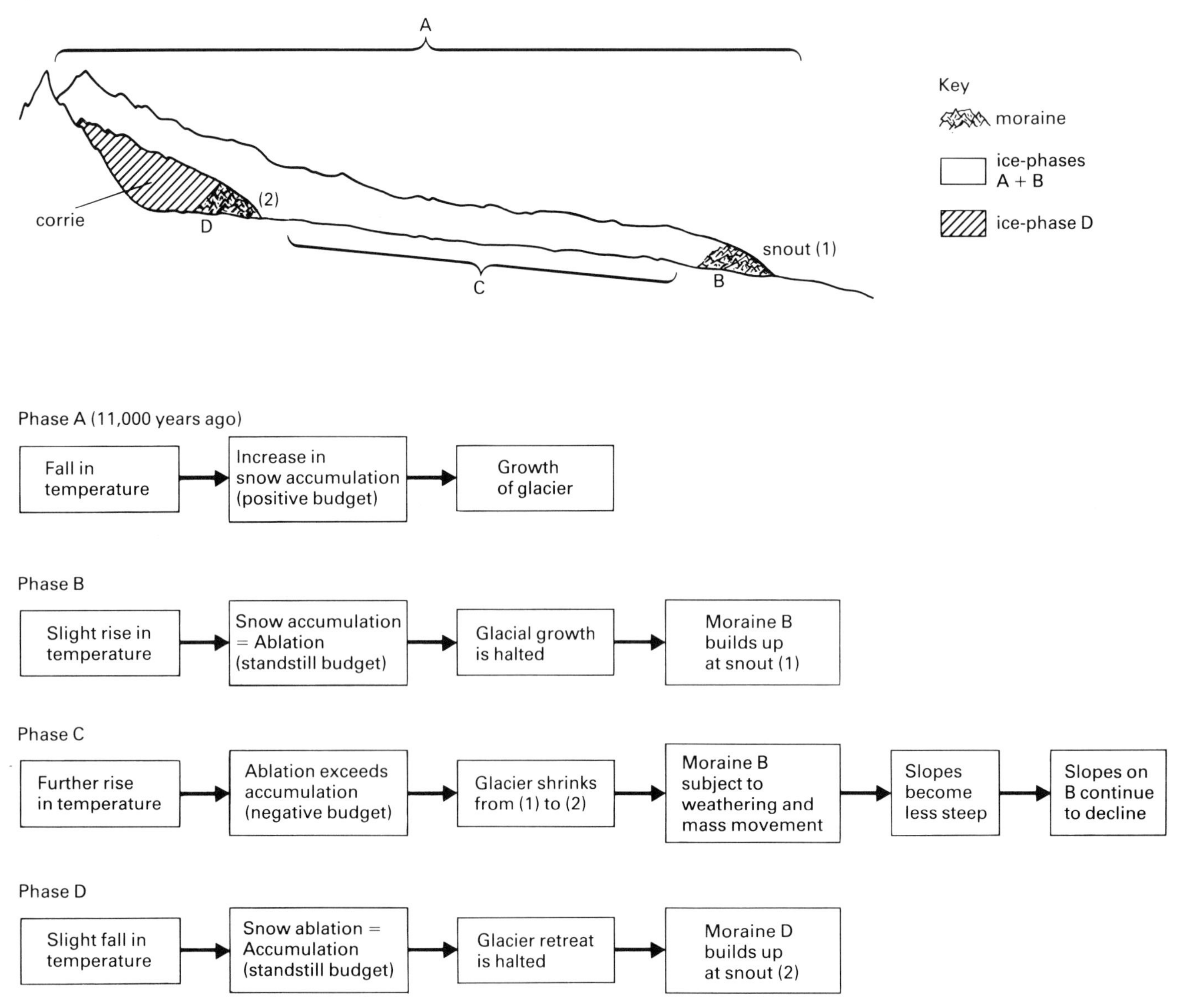

14.2 Moraine shape

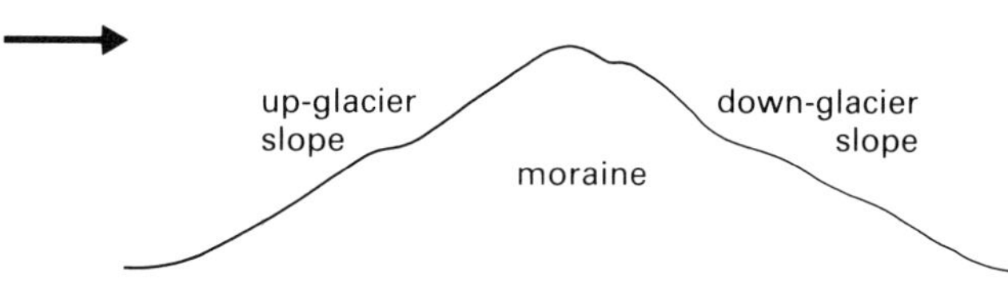

14.3 Distribution of moraines in a valley

Possible hypotheses

W L Graf carried out a study of moraines in Montana and Wyoming, USA. He found that the newest moraines had the steepest slope and that the up-glacier and down-glacier slopes of moraines are generally fairly symmetrical (figs. 14.2 and 14.3).

Before we set out hypotheses it is necessary to make a basic assumption – that the newest moraines are nearest the head of the valley or corrie. It is now possible to see whether Graf's findings apply to moraines in Britain.

1 The newer the moraines the steeper the angle.
2 The angles of slope on the up-glacier and down-glacier sides of the moraines are symmetrical.

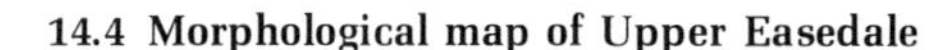

14.4 Morphological map of Upper Easedale

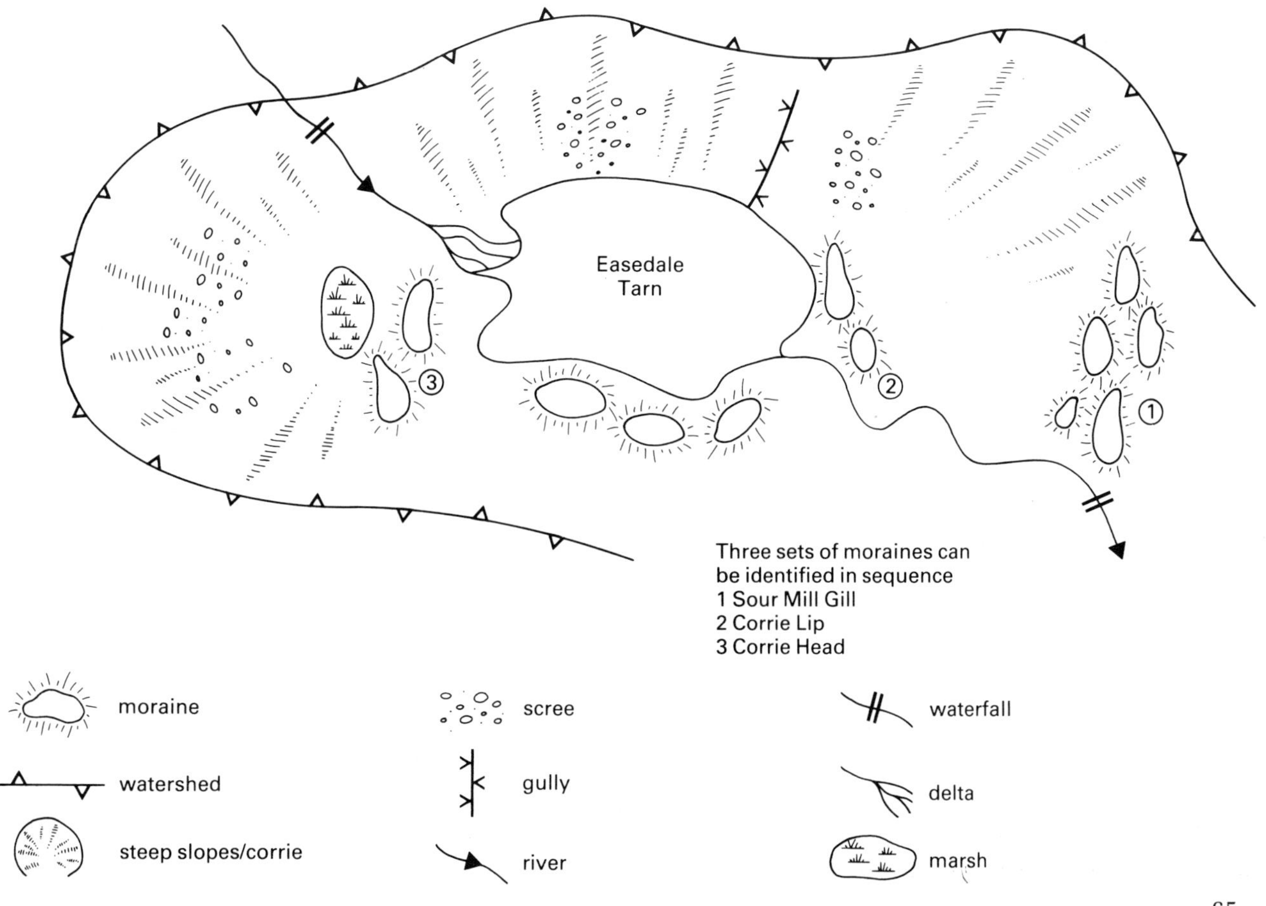

Location

Only attempt this study in an area with which you are familiar. You need to choose a fairly accessible glaciated valley which has a sequence of moraines. If possible choose one with a corrie at its head. Easedale in the Lake District is a suitable location, or alternatively the Cwm Glas/Llyn Llydaw/Cwm Dyli side of Snowdon.

Careful study of an OS map together with a relevant geology map showing surface deposits should help you to select a location. In the field moraines are usually readily distinguishable. They are often bracken-covered in contrast to their surroundings which may be marshy or grassy. Ask for advice at this stage.

Data collection

Advance preparation

Visit the area and construct a morphological map (fig. 14.4) to show the moraines you propose to study in the context of the valley as a whole. A morphological map records the main features of the landscape. (For morphological map symbols see Waters 'Morphological Mapping'.) Your map should indicate which are the newest and oldest moraines. You should now prepare a recording sheet (fig. 14.5) for each set you propose to measure.

Equipment

30-metre tape
clinometer
ranging rods if working alone
compass

You should do at least ten transects from the upstream to downstream slopes on each set of moraines in order to collect sufficient data. Set out your recording sheet as shown in fig. 14.5.

14.5 Moraine recording sheet

Location ____________												
Transect	1				2				3			
	D	F	B	A	D	F	B	A	D	F	B	A
Crest etc.	________											

D = distance (metres)
F = foresight
B = backsight
A = mean angle
+ / − { record whether you are proceeding up or downslope next to the mean angle as shown in fig. 14.11.

Be sure to indicate the *crest* at each moraine by ruling across the transect as shown when you are actually measuring.

Recording

End moraines are often very complex and consist of several distinct mounds or ridges. In making your ten transects you should attempt to survey the central portion (known as the *width of the minor semi axis*) on ten separate morainic mounds. If this is impractical you may make more than one transect parallel to the short axis of the same mound. Beware of taking all your readings across one mound as it may be untypical of the morainic group as a whole.

In fig. 14.6 there are only eight mounds so you would need to do *two* transects on two of the mounds. You would need to use your compass to make sure that both transects were parallel to the minor semi axis. Make two transects of the two largest mounds. See chapter 6 for the method of slope recording. Remember to record whether you are going up or downslope (+ or −) next to the mean angle as shown in fig. 14.11.

Repeat the exercise at each set of end moraines you propose to measure. In Easedale the exercise could be carried out on the three sets of moraines shown on the map. However it would be possible to compare two sets only.

Processing data

1 Underline the steepest angle in each transect.
2 Calculate the mean angle in each transect.
3 Set out your mean and steepest angle results in a combined results sheet (fig. 14.7).

14.6 Moraine mounds forming an end moraine

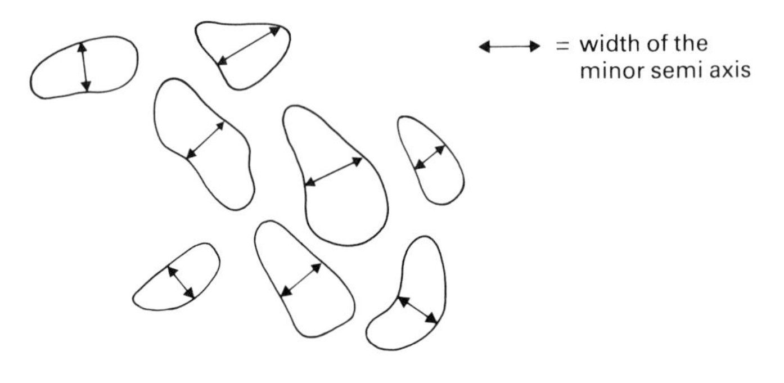

14.7 Combined results sheet for Easedale Moraines

	Sour Milk Gill		Corrie Lip		Corrie Head	
Transect	Mean angle (°)	Steepest angle (°)	Mean angle (°)	Steepest angle (°)	Mean angle (°)	Steepest angle (°)
1	16	18	14	32	20	29
2	16	21	11	22	14	33
3	5	15	16	25	24	35
4	11	15	17	23	14	27
5	5	12	11	19	18	29
6	11	19	12	17	17	30
7	8	24	23	28	15	30
8	6	16	22	29	12	24
9	8	17	14	24	15	26
10	6	13	8	11	16	25

Presentation of results

1 Draw one set of dispersal diagrams to show the steepest angles and another set to show the mean angles.
Dispersal diagrams are used to represent quantified (e.g. angle) and non-quantified (e.g. locality) data.
Now calculate the median for each location on each graph. (The median is the midpoint on a ranked list or series of plotted points, i.e. with n points it is the $(n+1)/2$ point. Draw a line across each column to represent the median. Note that in fig. 14.8 n (the number of points) $= 10$. $(10+1)/2 = 5.5$ so your line will be drawn midway between points 5 and 6. Should points lie on the same *value* (see mean angle for Sour Milk Gill) draw the line *through* both points.

14.8 Dispersal diagrams showing angles of slopes at different localities

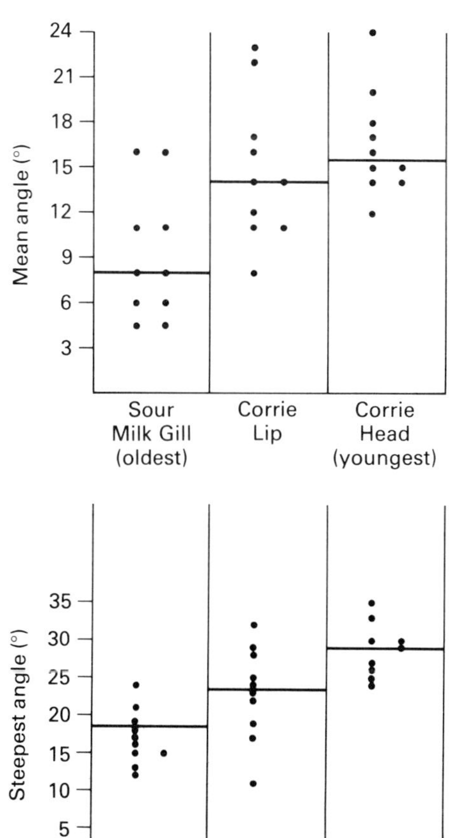

14.9 Matrix showing mean angles of slopes

Location	Mean angle (°) 0–9	10–19	20–29	30–39
Sour Milk Gill	6	4	0	0
Corrie Lip	1	7	2	0
Corrie Head	0	8	2	0

Now repeat for steepest angles.

2 Alternatively, you may set out a *matrix* (fig. 14.9) to show the mean and steepest angles. You will need to *group* the angles into classes.
Now repeat for steepest angles.

3 For *each* location, e.g. Sour Milk Gill, draw a set of ten superimposed profiles (see chapter 11). Choose a suitable scale for your data.

4 Complete a matrix showing the up-glacier and down-glacier angles for each set of moraines, using your original data for each transect. If you started measuring from the up-glacier side the up-glacier angles will usually be shown as positive and the down-glacier as negative, but this may not always be the case (fig. 14.11).

5 Now use your data to complete a *frequency polygon* (fig. 14.13). Put the *class frequency* on the y axis and the *class mark* on the x axis. The class mark is the midpoint of the class; thus in the class 0–4 the class mark is 2.

14.10 Superimposed moraine profiles

14.11 Angles of one transect

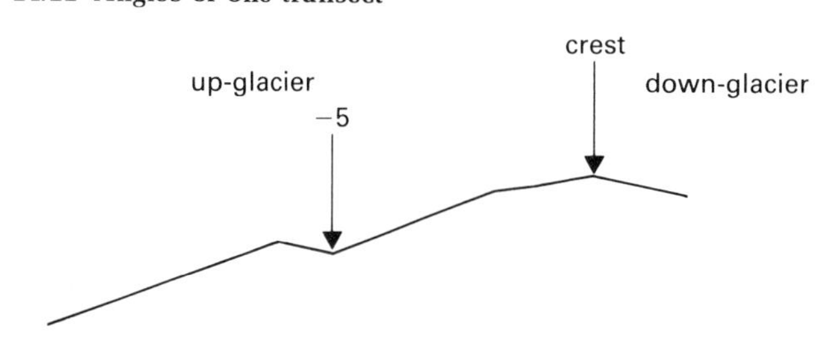

	Distance	F	B	A	
	20	15	15	+ 15	Up-glacier
	5	6	4	− 5	
	25	16	16	+ 16	
	4	3	3	+ 3	
	4	7	5	+ 6	
	4	6	4	+ 5	
Crest					
	11	10	10	− 10	Down-glacier

14.12 Complete matrix of frequency of angles for Sour Milk Gill

	Recorded angles 0°–4°	5°–9°	10°–14°	15°–19°	20°–24°	Total
Up-glacier	20	13	13	17	1	64
Down-glacier	16	13	11	13	0	53
Total	36	26	24	30	1	117

14.13 **Frequency polygon for angles of slope**

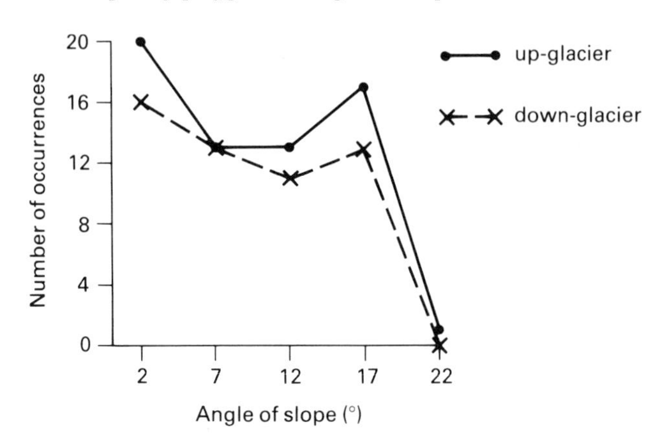

Analysis

Hypothesis 1

Use your dispersal diagrams to help you investigate hypothesis 1. Describe the *distribution* of the data about the median and the *range* of values for each location. Compare the distribution and ranges for all localities. If there is an obvious difference in the median value, distribution of points and ranges of points for each location you may accept your hypothesis, provided that the median values increase from oldest to youngest moraines (see fig. 14.8). If you have used a matrix to analyse your data it should also show that steeper angles are found on newer moraines if your hypothesis is correct. You must now attempt to explain your results with reference to the introduction and standard glacial texts.

If you do not find a pattern similar to that shown in fig. 14.8 you will have to consider what factors other than age determine the steepness of moraines. The angle of rest of the particles comprising the moraine may be very significant. Some moraines may be thinly plastered over solid rock, thus creating steep angles. (See suggestions for further work.)

Hypothesis 2

Your superimposed profiles should give you some visual information about the symmetry of each moraine mound and also of the overall symmetry of each set of moraines. The frequency polygon will confirm this visual impression. If the up-glacier and down-glacier lines are close together on your polygon you can accept your hypothesis. Try to explain why symmetry occurs if you have been able to accept your hypothesis; conversely, try to account for any lack of symmetry if your hypothesis has been rejected (e.g. a river may be under cutting the down-glacier slope of a moraine, causing slumping and/or rapid soil creep, and producing steeper angles than on the up-glacier slope).

Suggestions for further work

A particle size analysis of each set of moraines could be made. Particle orientation (showing the direction of glacial movement) could also be undertaken. See Hanwell and Newson *Techniques in Physical Geography*, chapter 8 C3.

You could compare moraines in similar locations in adjacent valleys.

Further reading

C Embleton and C A M King, *Glacial Geomorphology*, Arnold 1975

J D Hanwell and M D Newson, *Techniques in Physical Geography*, Macmillan 1973

R S Waters, 'Morphological Mapping', *Geography* 43 (1958) 10–17

Appendix A Sampling of areal distributions

It is often impossible or impractical to collect a complete set of information about a particular distribution. Therefore it is necessary to take a sample of the total population. The sample is usually referred to as *n*, the total population is usually referred to as *N*.

It is essential to collect a sample which is representative of the total population, thus it must not be too small and it must avoid bias. There are three commonly used methods of sampling.

1 Random sampling

This is the only method where every element of the population has an equal chance of selection. Tables of random numbers are usually used to generate numbers which are then used to select the sample. If the study requires points over an area a grid system must be constructed over a map of the area and the random numbers then used as grid references (see fig. A1). Alternatively if the data are in a list, e.g. factories, the random numbers can be used to select items from the list.

2 Systematic sampling

This is usually quicker than random sampling and is often equally suitable. However, it is possible to select a regular variation which would thus be over emphasised in the sample. Where the sample required is from an area then a series of points that are equally spaced should be produced (see fig. A2). If the data are in a list then the information should be taken regularly, every sixth item for example.

3 Stratified sampling

This is appropriate where the data are divided into different sections or strata. For example an area may be divided into different regions according to rock type; or data about a human population may be divided up into age groups. Having identified the strata the sample may then be taken by random or systematic methods (see fig. A3).

Further reading

R Dougherty, *Data Collection*, OUP 1974

P Toyne and P. Newby, *Techniques in Human Geography*, Macmillan 1971

A1 Random sampling: points generated by random number tables

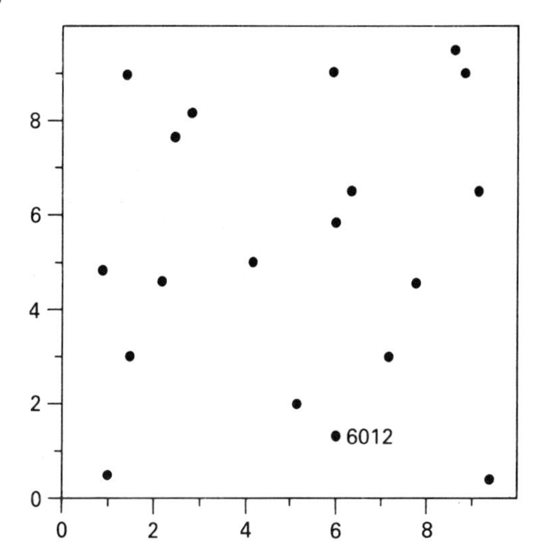

A2 Systematic sampling

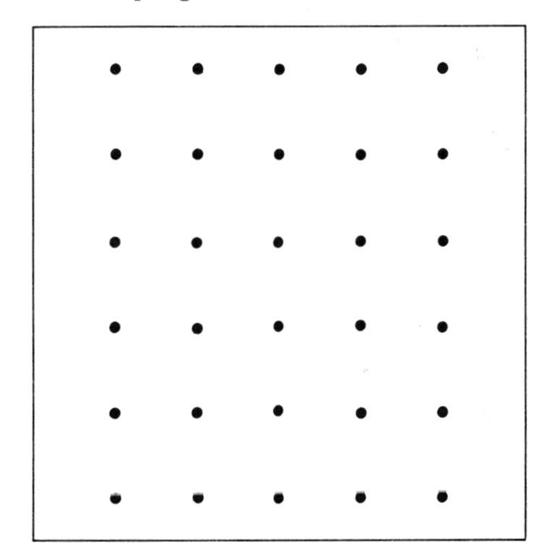

A3 Stratified sampling: random samples within two strata. The number of points within each region should be proportional to the size of that region.

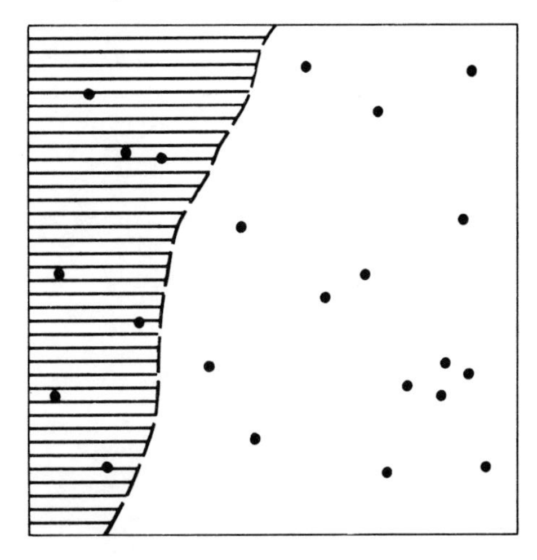

Appendix B Random number table

89 90	26 36	22 74	71 13	74 05	58 67	79 59	34 69	67 51	58 57	76 58	17 38
86 38	25 76	20 69	35 84	53 73	47 38	46 54	91 63	33 65	25 02	32 03	66 23
98 10	31 80	82 41	41 82	54 76	90 22	95 78	25 58	06 68	19 49	08 95	74 83
94 18	87 43	71 84	45 45	96 43	59 63	23 13	54 79	61 67	36 37	33 26	87 75
33 69	26 21	93 49	56 64	25 68	49 58	90 47	33 17	19 56	33 36	11 72	19 09
85 71	59 36	22 42	88 11	63 09	63 24	95 76	07 89	89 07	73 09	74 88	23 55
32 25	01 53	36 19	10 42	49 50	98 75	38 88	65 44	00 45	88 37	41 62	84 72
17 85	78 43	44 60	69 05	86 59	66 98	92 56	79 88	70 45	05 62	17 74	81 88
04 50	22 00	74 57	49 82	75 34	25 90	05 50	24 63	30 64	31 84	66 19	83 47
79 91	04 95	34 45	08 62	83 99	81 26	89 40	61 49	94 30	94 66	08 26	10 42
48 01	83 62	70 23	81 29	23 84	38 91	91 50	19 93	25 54	61 74	26 05	34 65
95 14	63 02	09 68	89 66	32 52	16 05	39 39	02 71	55 99	16 45	98 67	01 58
58 51	58 85	13 72	52 57	84 58	66 68	32 58	66 07	78 74	03 42	81 93	73 32
88 50	46 86	70 24	86 62	38 78	33 76	68 20	67 61	29 72	52 17	44 56	96 53
16 54	05 64	73 80	20 83	44 09	66 51	28 72	97 48	68 60	20 96	55 47	55 73
20 73	33 39	25 44	19 13	68 71	26 81	25 14	30 75	74 61	56 70	93 47	12 50
47 38	74 44	92 47	37 84	03 89	26 23	52 41	06 59	38 43	03 46	88 95	91 96
58 17	60 57	02 99	45 72	45 42	61 47	15 18	73 12	98 26	83 31	83 99	24 07
26 81	50 42	24 97	87 14	89 71	83 17	14 93	46 12	16 74	95 10	99 87	21 05
95 64	20 88	42 63	17 25	80 41	33 22	25 83	65 08	33 05	89 61	29 86	12 19
29 80	09 62	33 60	64 44	31 22	31 70	18 11	07 85	70 74	91 17	91 45	76 77
70 46	91 35	07 77	27 84	40 20	77 25	44 00	03 23	45 60	02 22	61 67	08 49
96 67	06 01	60 64	69 46	17 69	27 55	98 32	05 66	88 46	68 25	87 25	14 16
12 25	05 68	84 84	62 91	23 97	99 19	93 94	66 51	22 45	41 59	66 80	62 63
18 65	57 35	74 25	81 35	62 62	01 43	74 05	88 51	53 69	88 78	15 71	96 11
52 49	40 62	32 80	98 03	42 14	61 02	77 11	87 66	70 56	20 35	02 91	51 30
13 62	74 91	12 44	96 86	45 98	55 33	06 71	98 91	20 22	29 56	66 82	80 02
33 31	40 35	41 32	48 21	85 70	71 89	76 31	87 22	55 58	90 72	39 82	53 69
80 77	73 73	18 15	96 48	26 00	60 98	89 18	07 71	62 39	86 97	15 97	09 82
69 82	64 04	67 28	95 36	60 92	32 00	68 83	62 23	32 55	61 16	94 01	09 82
31 08	33 58	06 53	24 59	05 09	42 83	67 73	48 81	85 62	92 49	25 93	56 45
58 11	60 05	86 12	24 52	89 45	93 31	15 87	94 84	90 03	64 06	15 63	01 73
68 32	40 38	37 45	39 99	26 02	59 97	12 64	67 43	86 80	14 45	60 08	12 08
22 45	63 38	51 44	94 68	48 49	56 06	09 22	52 95	50 13	84 46	80 27	65 93
00 38	56 12	37 21	19 81	02 72	75 49	44 41	24 66	55 22	60 79	80 97	83 06
84 76	13 46	00 67	48 58	42 75	77 04	97 19	66 07	55 66	38 61	68 17	66 99
97 80	90 05	58 62	97 03	58 03	05 59	38 74	54 95	87 04	59 87	81 07	67 94
26 86	92 06	77 04	10 51	91 26	24 43	94 89	39 62	88 72	11 52	66 47	51 79
39 24	80 25	00 41	97 85	59 58	63 97	89 09	22 21	11 86	51 18	33 08	95 67
66 58	86 48	27 66	78 53	22 51	25 31	87 95	74 75	50 04	42 05	71 92	13 15
73 64	76 36	02 06	33 96	35 22	85 86	96 30	00 18	05 37	41 90	85 50	99 13
85 23	58 50	31 15	30 82	44 92	72 55	58 59	98 92	39 34	60 94	48 68	85 58
53 10	34 43	02 07	65 97	57 25	39 09	15 69	67 79	91 75	71 27	95 74	86 20
53 58	40 68	34 29	94 10	07 42	07 22	17 24	36 42	59 05	70 19	07 71	88 72
45 40	02 88	03 70	75 11	91 74	11 68	56 39	86 08	00 73	41 56	39 75	89 50
61 72	39 11	34 39	68 80	08 81	64 23	74 68	28 94	02 59	82 22	55 39	51 19
84 30	20 34	25 31	26 51	65 21	77 37	79 33	53 47	79 02	51 78	79 17	66 05
34 07	51 42	92 86	47 09	52 90	94 27	47 73	55 39	38 62	93 20	92 75	38 53
20 59	05 44	83 29	01 98	88 06	51 48	45 38	63 98	41 93	87 79	18 13	31 90
29 87	94 10	13 43	59 63	45 98	29 97	26 55	46 21	42 55	02 07	06 05	95 88

Appendix C Significance levels for statistical tests

Critical values of ρ, the Spearman rank correlation coefficient. Rho must be equal to or more than the stated value to be significant.

n	Significance level	
	.05	.01
10	.564	.746
12	.506	.712
14	.456	.645
16	.425	.601
18	.399	.564
20	.377	.534
22	.359	.508
24	.343	.485
26	.329	.465
28	.317	.448
30	.306	.432

Critical values of χ^2. χ^2 must be equal to or more than the stated value to be significant.

df	Significance level		
	0.05	0.01	0.001
1	3.84	6.63	10.83
2	5.99	9.21	13.81
3	7.81	11.34	16.27
4	9.49	13.28	18.47
5	11.07	15.09	20.52
6	12.59	16.81	22.46
7	14.07	18.48	24.32
8	15.51	20.09	26.12
9	16.92	21.67	27.88
10	18.31	23.21	29.59
11	19.68	24.73	31.26
12	21.03	26.22	32.91
13	22.36	27.69	34.53
14	23.68	29.14	36.12
15	25.00	30.58	37.70
16	26.30	32.00	39.25
17	27.59	33.41	40.79
18	28.87	34.81	42.31
19	30.14	36.19	43.82
20	31.41	37.57	45.31
21	32.67	38.93	46.80
22	33.92	40.29	48.27
23	35.17	41.64	49.73
24	36.42	42.98	51.18
25	37.65	44.31	52.62
26	38.89	45.64	54.05
27	40.11	46.96	55.48
28	41.34	48.28	56.89
29	42.56	49.59	58.30
30	43.77	50.89	59.70
40	55.76	63.69	73.40
50	67.50	76.15	86.66
60	79.08	88.38	99.61
70	90.53	100.4	112.3
80	101.9	112.3	124.8
90	113.1	124.1	137.2
100	124.3	135.8	149.4